T0391185

Lactoferrin and its Role in Wound Healing

Yoshiharu Takayama

Lactoferrin and its Role in Wound Healing

 Springer

Yoshiharu Takayama
Functional Bio-molecules
Research Team
National Institute of Livestock
and Grassland Science
Ikenodai 2
305-0901 Tsukuba, Ibaraki
Japan
takay@affrc.go.jp

ISBN 978-94-007-2466-2   e-ISBN 978-94-007-2467-9
DOI 10.1007/978-94-007-2467-9
Springer Dordrecht Heidelberg London New York

Library of Congress Control Number: 2011939477

Printed on acid-free paper

Springer is part of Springer Science+Business Media (www.springer.com)

I dedicate this book to my father, Noboru Takayama, who passed away during the writing of this book.

Preface

Lactoferrin (lactotransferrin) was first identified as an iron-binding protein abundantly found in milk. It is synthesized by glandular epithelial cells and secreted into body fluids such as saliva, tears and mucosal secretions. Lactoferrin has been considered to play important roles in host defense, since lactoferrin shows broad spectrum of anti-bacterial and anti-viral activities, arising from its iron-chelating property. However, some functions of lactoferrin are independent from its iron chelating ability.

In addition, lactoferrin is known as a major component of secondary granules of neutrophils. Plasma lactoferrin concentration is dramatically increased by bacterial infection or tissue injury. Lactoferrin can regulate the function of innate and adaptive immune cells and exhibits immunoregulation activity.

One of the most novel activity described for lactoferrin is its regulatory function in wound healing. The results of animal studies indicate that the topical administration of lactoferrin enhances the rate of skin wound closure in control and diabetic mice. Skin wound healing is a complex physiological process that requires the regulation of different types of cells such as immune cells, keratinocytes, fibroblasts, and endo-thelial cells. Inflammation is prerequisite for wound healing. The promoting effect of lactoferrin on wound closure is likely to dependent on its immune regulatory function. In addition, the results of recent *in vitro* and *in vivo* studies show that lac-toferrin is a potent regulator of dermal fibroblasts and epidermal keratinocytes, and promotes skin or corneal epithelial wound healing by increasing their proliferation, migration or deposition of extracellular matrix (ECM) components such as collagen and hyaluronan. Hyaluronan (hyaluronic acid) is a glycosaminoglycan distributed throughout connective and epithelial tissues. The role of hyaluronan in wound heal-ing is the promotion of cell proliferation and motility.

The aim of this book is to describe how lactoferrin directly or indirectly regulates the wound healing processes that consists of multiple steps. Lactoferrin exerts its biological effects by binding to specific lactoferrin receptors on target cells. The diverse functions result from activation of different receptors and signal transduction path-ways, and also from differences in the expression levels of receptors. This book will describe an overview of molecular basis of wound healing (Chap. 1), in addition to providing a general review of hyaluronan (Chap. 2) and lactoferrin (Chap. 3) and role of lactoferrin as signaling mediator (Chap. 4). Finally, the effects of lactoferrin on wound healing are summarized in Chap. 5.

The multi-functionality of lactoferrin fosters usage of lactoferrin as a therapeutic agent for non-healing chronic wounds. A special feature of chronic wound is prolonged or excessive inflammatory response at the wounded site. Current therapy using recombinant growth factor is partially effective for healing of chronic wounds. In the clinical trial of patients with diabetic foot ulcers, the group treated with human recombinant lactoferrin showed significant improvement of wound healing rates compared with placebo group. Based on these findings, lactoferrin should be used in patients with diabetic and other types of ulcers.

Yoshiharu Takayama

Contents

Chapter 1
Molecular Regulation of Skin Wound Healing

Abstract The skin wound healing is complex physiological event for restoration of the intact structure in injured tissues. It begins with hemostasis and is followed by inflammation, which is prerequisite for subsequent events such as reepithelialization, granulation tissue formation, and wound contraction. Reepithelialization and granulation tissue formation in turn involves proliferation and migration of keratinocytes and fibroblasts, respectively. Wound contraction contributes for the reduction of the wound size and therefore to shorten the healing period. Many growth factors, cytokines, chemokines, and proteases regulate cell functions during the wound healing process. Spatial and temporal alterations in the actions of these molecules result in the failure of wound healing, non-healing chronic wounds.

Keywords Chronic wounds • Granulation tissue formation • Inflammation • Reepithelialization • Wound contraction

1.1 Introduction

Adult skin consists of two tissue layers, epidermis and underlying dermis (Fig. 1.1). Epidermis is composed of stratified layers of keratinocytes (basal, subbasal, spinous, granular and cornified layers). The differentiation of keratinocytes toward a terminally differentiated corneocyte is very tightly regulated. Basal keratinocytes divide with daughter cells, migrating into the overlying spinous layer. Keratinocytes in the spinous layer subsequently move into a granular layer, and eventually move into the outer cornified layer of the epidermis. Dermis is thick layer of collagen rich connective tissue, consisting with fibroblasts, nerves, capillaries and lymphatic vessels. Fibroblasts are ubiquitous and a predominant cells in connective tissue. They are responsible for deposition of extra cellular matrix (ECM) and matrix remodeling. Epidermis and dermis are separated by the basement membrane, a sheet composed of specialized collagens and matrix proteins such as laminins and collagen IV.

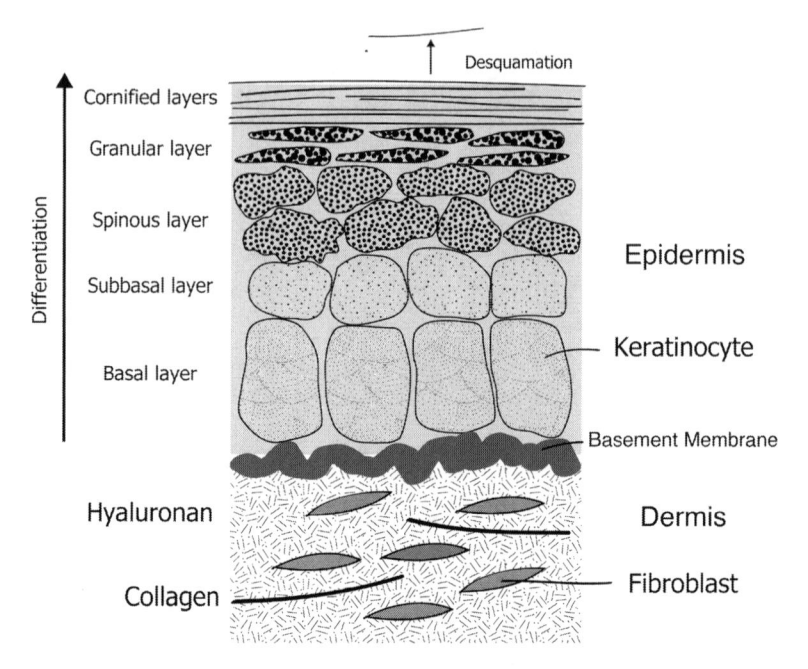

Fig. 1.1 Structure model of dermis and epidermis. Epidermis consists of layer of keratinocytes. The proliferating keratinocytes are located in the basal layer. During the process of differentiation, the basal cells lose their proliferative potential and migrate toward the surface of the epidermis. Dermis is composed of fibroblasts and extracellular matrix (ECM) components, which secreted from the fibroblasts

The skin acts as a protective barrier to the outside environment. Accordingly, any break in the skin has to be repaired rapidly to prevent blood loss and to close gaps that expose the tissue to the environment. Subsequent wound healing is an essential process for restoration of the intact structure in injured tissue, encompassing a number of overlapping phases, including inflammation, reepithelialization, wound contraction, angiogenesis, scar formation and remodeling (Fig. 1.2). They are tightly regulated by numerous growth factors, cytokines and proteases. Although elimination of important mediator may be compensated by redundant mechanisms, deletion of the critical regulators results in pathogenic and chronic wound healing. In this chapter, the ability of the mediators to regulate cellular proliferation, migration and adhesion is summarized on the basis of both loss- and gain-of function experiments.

1.2   Immediate Response

Tissue injury causes the immediate response to injury. Wounds cause leakage of blood from damaged blood vessels. The formation of plasma clot is necessary to stop local hemorrhages immediately. It is initiated by proteolytic conversion of

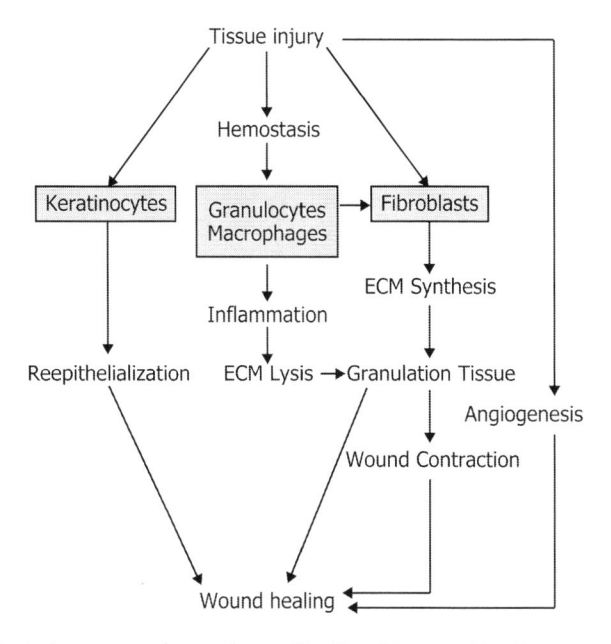

Fig. 1.2 Biological processes of normal wound healing. The wound healing process is composed of sequential events, and begins with hemostasis and inflammation. Wound healing involves reepithelialization, granulation tissue formation and wound contraction. Reepithelialization and granulation tissue formation requires proliferation and migration of keratinocyres and fibroblasts, respectively. Wound contraction contributes for the reduction of the wound size and therefore to shorten the healing period

fibrinogen into fibrin by thrombin. As fibrin molecules assemble into fibrin fibers, platelets and neutrophils are entrapped in a mesh of fibrin fibers. It act as a temporally shield protecting the denuded tissue [1, 2]. The process of clotting induces platelet degranulation and release of cytokines and growth factors, including platelet-derived growth factor (PDGF), insulin-like growth factor-1 (IGF-1), epidermal growth factor (EGF) and transforming growth factor-β (TGF-β) [3]. Plasma clot is served as a provisional matrix for invading cells as well as a reservoir of cytokines and growth factors that are released from activated platelets [3]. These growth factors act as chemotactic factors for lymphocytes, fibroblasts and keratinocytes, and promote various processes of reepithelialization and wound contraction.

1.3   Inflammatory Response

The inflammatory phase begins with the infiltration of leukocytes (mainly neutrophils) from damaged blood vessels into wounded sites. The infiltration of leukocytes is accompanied by activation of existing immune cells, such as mast cells

[4], δγ T cells [5], and Langerhans cells [6]. They are attracted by chemotactic signals released from activated platelets, trapped in plasma clot. Chemokines were first identified as proteins that are critical to the process of inflammation. Monocyte chemotactic protein-1 (MCP-1/CCL2) acts as strong chemotactic signal for macrophages and monocytes [3]. Elevation of MCP-1 mRNA is observed in both human and mouse wound healing models [7–9]. The MCP-1 knockout mice display delayed wound closure [10]. IL-8 (CXCL8), GROα (CXCL1) and IP-10 (CXCL10) are additional chemokines with lymphocyte-attractant properties [3]. Furthermore, peptides cleaved from bacterial surface proteins, ECM and fibrin act as chemotactic signals for the leukocytes. At the wounded site, P-selectin and E-selectin are induced in the surface of endothelial cells. They induce leukocyte activation and adhesion to endothelial cells. Wound closure is delayed in P-selectin and E-selectin double knockout mice or I-CAM knockout mice due to reduced infiltration of neutrophils and macrophages into wounded site [11, 12]. The activated leukocytes crawl out between endothelial cells into extravascular space. Neutrophils arrive at the wounded site within minutes of injury, and kill contaminating bacteria by phagocytosis as well as production of reactive oxygen species (ROS). In addition, they act as a source of pro-inflammatory cytokines (Tumor necrosis factor-α (TNF-α), IL-1, and IL-6) to activate fibroblasts and keratinocytes [13]. Unless a wound is infected, the infiltration of neutrophils is stopped after few days.

Monocytes are recruited from the circulation later than neutrophils. They reach peak a day after injury [14]. Monocytes rapidly differentiate into macrophages after migration from the vasculature to the wound site [15]. The task of macrophages is phagocytosis of remaining micro-organisms, matrix debris and cells, including neutrophils [16]. The clearance of debris is prerequisite for the resolution of inflammation. In addition to pro-inflammatory cytokines, the activated macrophage releases growth factors such as EGF, fibroblast growth factor-2 (FGF-2), PDGF, vascular endothelial growth factor (VEGF), TGF-α and TGF-β [17]. Inhibition of macrophage infiltration results in impaired wound healing [18].

Mast cells are important source of pro-inflammatory cytokines and chemokines. They are present in the skin and transiently activated in response to injury. In mast cell-deficient mice, infiltration of neutrophils into a wounded site is impaired, whereas infiltration of macrophage is not affected, suggesting that mast cells are involved in recruitment of neutrophils [19]. Mast cells release histamine and regulate vascular permeability.

δγ T cells are dendritic epidermal T-cells. They recognize antigen expressed by epidermal keratinocytes. Deficiency of δγ T cells results in delayed wound healing in response to mechanical injury. δγ T cells also act as an important source of keratinocyte growth factor-1 (KGF-1/FGF-7), KGF-2 (FGF-10) and IGF-1 [20]. KGF promotes kerarinocyte mediated hyaluronan deposition in wound area [21].

1.4   Reepithelialization

Keratinocytes are epithelial cells and participate in the body's first line of defense to the outside environment. In response to injury, keratinocytes in the basal and sub-basal layers of the dermis begin to proliferate and then migrate into wounded area until a new epithelium cover the damaged area (Fig. 1.3) [1, 22]. The conversion of keratinocytes to a migratory phenotype is designated as Epithelial Mesenchymal Transition (EMT), in which there is a loss of cell adhesion, down-regulation of E-cadherin expression, and increased cell mobility [23]. IL-1 is released from the keratinocytes, in an autocrine fashion [24]. Furthermore, IL-1 also acts as alert signal for surrounding tissue leads to chemotaxis of neutrophils and macrophages, proliferation of fibroblasts. The keratinocyte proliferation is also mediated by local release of growth factors, including TNF-α, EGF, heparin-binding EGF-like growth factor (HB-EGF) and KGF-1 [25–29]. The migration of keratinocytes is independent from its proliferation. The rate of migration is regulated by the tissue oxygen tension and the humidity of the environment. Members of EGF family are potential regulators for keratinocyte migration [27, 30]. TGF-α is known as most potent inducer of keratinocyte migration among EGF receptor (EGFR) ligands [31]. HB-EGF is secreted by keratinocytes at the leading edge of reepithelialization in an autocrine mechanism [32]. These results are consistent with the study using animal model. Administration of EGF or TGF-α promotes the reepithelialization of burn wounds on the backs of pigs [33, 34]. Despite of a possible functional redundancy among EGF family members, reepithelialization is impaired in TGF α [35] and HB EGF knockout mice [36].

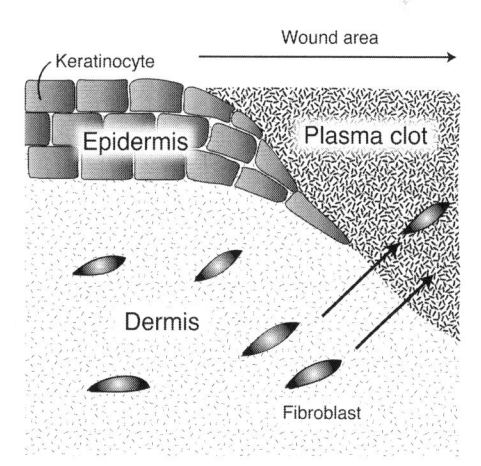

Fig. 1.3 Schematic drawing of granulation tissue formation and reepithelialization. Wound fibroblasts are derived from cells surrounding the wound. They are responsible for replacing the plasma clot (fibrin clot) with collagen-rich granulation tissue. Reepithelialization after skin wounding requires both migration and proliferation of keratinocytes in proximity of the wound

FGF-2 promotes keratinocyte migration and reepithelialization of human skin explants [37, 38]. Wound reepithelialization is enhanced in exogenous administration of FGF-2 in animal model [39, 40]. FGF-2-deficint mice show delayed wound healing [41, 42]. These lines of observations indicate essential role of FGF-2 in reepithelialization and wound healing.

KGF-1 is mainly provided from dermal fibroblasts and γδ T-cells. KGF is upregulated more than 100-fold at a wounded area in response to pro-inflammatory cytokines [43]. Decreased KGF expression results in impaired wound healing in glucocorticoid treated mice and genetically diabetic mice [44]. KGF prevents ROS-induced apoptosis of keratinocytes by increasing transcription factors involved in detoxification of ROS. Expression of dominant negative FGF receptor 2 (FGFR-2) in basal keratinocytes blocks their responsiveness for KGF-1 and reepithelialization, suggesting the important role of KGF-1 [27]. However, reepithelialization defect is not observed in KGF-1-deficient mice, due to compensatory effect of KGF-2, which also secreted from dermal fibroblasts and γδ T-cells [45].

TGF-β is an important regulator in multiple steps of wound healing [3]. TGF-β is released from platelets, neutrophils, dermal fibroblasts and migrating keratinocytes. It is controversial whether TGF-β promotes or inhibits the reepithelialization. Indeed, TGF-β promotes keratinocyte migration, but inhibits keratinocyte proliferation. In Smad3-deficient mice, abrogation of TGF-β signaling correlates with accelerated reepithelialization [46]. TGF-β antagonist increases porcine full thickness wound reepithelialization [47]. These lines of observations suggest that TGF-β negatively regulates reepithelialization. On the other hand, TGF-β stimulates the expression of fibronectin and integrins toward a more migratory phenotype during reepithelialization. Overexpression of TGF-β stimulates reepithelialization of partial thickness wounds [28, 48], suggesting that promoting effects of TGF-β on the reepithelialization.

Very little is known about which signal pathway participates in promotion of keratinocyte migration. Classical MAPK pathway is likely involved in the regulation of keratinocyte motility. Hepatocyte growth factor (HGF) and KGF can activate ERK1/2 and p38 MAPK. Inhibition of either the ERK1/2 or p38 MAPK pathway results in delayed corneal epithelial wound healing [49]. Activation of EGFR signaling induces migration of keratinocytes during wound healing by MAPK-dependent pathway. The organization of actin cytoskeleton is controlled by Rho, Rac and Cdc42, members of small guanosine triphosphatase (GTPase). They are targeted to the membrane by posttranslational attachment of prenyl groups and cycle between an inactive GDP-bound form and an active GTP-bound form. Epidermal wound healing is apparently regulated by Rac1-dependent pathway. Overexpression of dominant-negative Rac1 or epidermis-specific deletion of Rac1 results in impaired epidermal wound healing [50]. Indeed, Rac1 participates in the keratinocyte migration by regulating the assembly of focal adhesion complex and actin cytoskeleton [51]. EGF-induced migration of keratinocytes is dependent on Rac1 signaling [52]. IGF-1 promotes keratinocyte migration, and indices membrane protrusion and facilitates cell spreading by Rho-dependent mechanism [53].

Epidermal keratin is a 40–70 kDa α-helical coiled-coil dimer of the intermediate filament. Specific pairs of type-I and type-II keratins are expressed in keratinocytes. Their expression profiles are dependent on maturation of keratinocytes [54]. Keratin filament links hemidesmosomes, nuclear matrix, and desmosomes within keratinocytes to impart physical strength to the epidermis. Reepithelialization is blocked in knockout mice lacking the bullous pemphigoid antigen (BPAG1), which links keratin filaments and hemidesmosome [55]. Basal keratinocytes of the stratified epithelium express keratins K5 and K14. Subbasal keratinocytes express keratins K1 and K10 [56]. In response to injury, the expression of K6, K16 and K17 keratins is increased [57]. IL-1, EGF, TGF-α and TNF-α can induce K6 and K16 expression in keratinocytes [58–60]. The physiological significance of the class switch of keratin is not clear. Overexpression of K16 in keratinocytes induces major changes in the organization of keratin filaments in a time- and calcium concentration-dependent manner [61]. It is hypothesized that K6 and K16 provides the plasticity and flexibility in migrating keratinocytes [62].

Integrins are heterodimeric trans-membrane receptor consisting of an α subunit and a β subunit that links the ECM to intracellular cytoskeleton. Ligand binding induces massive changes in cell shape and cytoskeletal organizations. The migrating keratinocytes change their integrin expression profile in order to crawl the provisional wound matrix and underlying wound dermis [63]. In intact skin, the basal keratinocytes constitutively express integrin $\alpha6\beta4$ and directly contact with the basement membrane [63]. $\alpha6\beta4$ is localized at the large adhesive complexes, called hemidesmosomes at the basal surface of keratinocytes [64]. The connection of $\alpha6\beta4$ to keratin filament is critical for integrity of hemidesmosomes. In addition, integrins sharing $\beta1$ subunit are found in clusters surrounding hemidesmosomes. They mediate cell attachment via large multi-molecular adhesion complexes known as focal adhesions and connects to the intracellular actin cytoskeletons. In migrating keratinocytes, hemidesmosome attachment to basement membrane is dissolved. The cytoplasmic tail of $\beta4$ is phosphorylated in response to HGF and EGF through the activity of protein kinase C (PKC) [65, 66]. The phosphorylation of $\alpha6\beta4$ integrin disrupts hemidesmosome structure [67, 68]. In migrating keratinocytes, expressions of integrins $\alpha5\beta1$, $\alpha v\beta5$, $\alpha v\beta6$, and $\alpha2\beta1$ are induced [69, 70]. They mediate cell adhesion to fibronectin ($\alpha5\beta1$, $\alpha v\beta6$), vitronectin ($\alpha v\beta5$) and collagen ($\alpha2\beta1$), and regulate signal transduction from the ECMs to the interior of cells [71–73]. The diversity of the integrin expression in migrating keratinocyte ensures cell adhesion to any component of the provisional matrix during reepithelialization [63]. The deletion of $\beta1$ integrins in keratinocytes causes a severe defect in wound healing, impaired migration along with hyperproliferative epithelium, suggesting a crucial role of $\beta1$ integrins in wound healing [74]. *In vivo* wound healing studies using integrin knockout mice indicate that reepithelialization is not dependent on specific types of integrins containing $\beta1$ subunit [63].

Once the wounded surface is covered with keratinocytes, the migration of epithermal is ended. The epidermal cells return to their original morphology and function [75]. Little is known about molecular mechanisms that involved in contact inhibition of keratinocytes.

1.5   Granulation Tissue Formation

In response to injury, fibroblasts in proximity of the wound migrate into the wounded clot (Fig. 1.3). The infiltration of fibroblasts starts at day 2 after injury and by day 4 they become a major cell type in a wounded clot. These cells synthesize ECM and replace the plasma clot with a collagen-rich matrix, called granulation tissue. PDGF and TGF-β promotes the formation of granulation tissue, since they promote the migration and collagen deposition of fibroblasts [3, 76–79]. TGF-β deficient mice show impaired late stage wound repair along with much thinner and less vascular granulation tissue [80]. Administration of TGF-β3 promotes formation of granulation tissue in a rabbit chronic wound model [81]. Connective tissue growth factor (CTGF) is also known to stimulate proliferation and chemotaxis of fibroblasts [82]. In addition, CTGF is strong inducer of collagen deposition by fibroblasts.

The functions of dermal fibroblasts are regulated by surrounding ECM as well as growth factors and cytokines. They bind to ECM through receptors belong to the integrin super family. Quiescent fibroblasts aligned with the collagen matrix express collagen binding integrin such as α2β1. When fibroblasts migrate into a plasma clot, they show different integrin expression profiles. Fibroblasts require fibronectin for migration from collagen matrix to plasma clots [22]. Fibronectin receptors are expressed by fibroblasts just prior to wound contraction [83]. The primary sequence motif of fibronectin for integrin binding is a tripeptide, Arg-Gly-Asp (RGD). The RGD motif is found in variety of adhesive proteins interacting with integrin receptors α3β1, α5β1, αvβ1, αvβ3 and αvβ5. The expression of integrins αvβ3 and α5β1 is increased in the migrating fibroblasts. When fibroblasts are cultured in a fibrin-fibronectin gel, PDGF induces expression of integrins α3β1 and α5β1, whereas PDGF treatment of fibroblasts cultured in collagen matrix increases α2β1 integrin expression [84]. Once the fibroblasts migrate into a wounded area, the expression of the RGD type integrin receptors is decreased, along with increased α2β1 expression.

1.6   Wound Contraction

During wound repair, fibroblasts develop mechanical stress to contract the wound [85]. Wound contraction contributes to the reduction of the wound size and therefore to shorten the healing period [1, 85]. Inhibition of fibroblast contraction abrogates the wound healing process [86, 87]. Fibroblasts cultured in a three-dimensional type I collagen gel are able to reorganize the surrounding collagen gel matrix into a more denser and more compact structure. This phenomenon, referred to as collagen gel contraction, is used as *in vitro* model of wound contraction. The contractility of fibroblasts occurs as a result of their motile activity [85, 88]. Many growth factors and cytokines, including PDGF, IGF, TGF-β, and thrombin promote the collagen gel contractile activity of fibroblasts [85, 89–92]. Adhesive interactions between fibroblasts and collagen are mainly mediated by β1 integrins, in particular α1β1 and

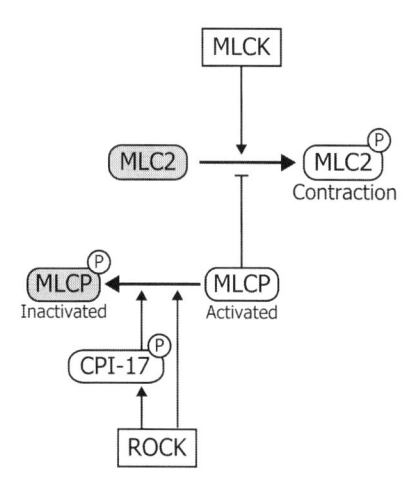

Fig. 1.4 Myosin light chain 2 (MLC2) phosphorylation is regulated by two distinct pathways. Myosin light chain kinase (MLCK) directly phosphorylates MLC2 and increases contractile activity of fibroblasts and myofibroblasts. Myosin light chain phosphatase (MLCP) reverses the effect of MLCK. Rho kinase (ROCK) increases MLC2 phosphorylation by phosphorylating (inactivating) MLCP

$\alpha2\beta1$ integrins. Several studies have demonstrated a role for the collagen binding integrins ($\alpha1\beta1$ and $\alpha2\beta1$) in collagen gel contraction by fibroblasts [93–96].

On the other hand, excessive wound contraction can result in formation of scar tissue [22]. Hence, fibroblast collagen contractility must be regulated in an optimal range to accomplish wound closure and also prevent scar tissue formation. Augmented PDGF production might be involved in the pathogenesis of scars and keloids [3]. Collagen gel contraction of fibroblast is inhibited by inflammatory cytokines such as IL-1, IL-8, interferon-γ (IFN-γ) and TNF-α [89, 90, 97, 98].

Fibroblasts have actin stress fibers that consist of bundles of actomyosin. The actin stress fibers are major mediators of fibroblast contractility [99]. Stress fiber contraction is dually regulated by a Ca^{2+}-dependent calmodulin/myosin light chain kinase (MLCK) system and a Ca^{2+}-independent Rho-kinase (ROCK/ROK) system (Fig. 1.4). Rho is a member of small GTPase and involved in the regulation of fibroblast motility. Microinjection of activated recombinant Rho into fibroblast results in rapid stress fiber formation [100]. ROCK is co-localized with actin stress fibers and is one of the downstream effecter of Rho [101, 102]. ROCK directly phosphorylates myosin light chain 2 (MLC2) at Ser19 [103]. The ROCK-induced MLC2 phosphorylation increases stress fiber contractility. Additionally, ROCK phosphorylates MBS subunit of myosin light chain phosphatase (MLCP) [99]. ROCK-induced MBS phosphorylation inhibits its phosphatase activity, and increases stress fiber formation by promoting MLC phosphorylation [104]. CPI-17 is an inhibitor of myosin phosphatase, and activation of CPI-17 leads to increased contractility. ROCK can phosphorylate and activate CPI-17 *in vitro* and in smooth muscle cells [105, 106]. However, it is not clear that CPI-17 is ROCK effecter in cells other than smooth muscle cells. In addition, zipper interacting protein kinase (ZIPK) directly phophorylates both MLC2 and MBS. These lines of observations indicate that ROCK promotes the stress fiber contractility by multiple mechanisms. On the other hand, MLC phosphorylation is also regulated by MLCK. MLCK directly phosphorylates MLC2 (Ser19) and increases myosin binding to actin. ROCK is necessary for maintaining the organization of the stress fibres and focal adhesions in the central portion of the cell. MLCK activity is required for maintaining their organization in the peripheral portion of the cell [107].

A part of fibroblasts are differentiated into myo-fibroblasts in granulation tissue. Myo-fibroblasts express α-smooth muscle actin (α-SMA) and show strong contractile activity as smooth muscle cells [108]. The fibroblast differentiation to myofibroblasts is enhanced by TGF-β and mechanical stimuli [109, 110].

1.7   Angiogenesis

Angiogenesis is essential for successful wound repair as nutrients and oxygen are provided by microvascular network in the granulation tissues. New capillaries become visible in wound bed 3–5 days after injury. It is started from the generation of new blood vessels, involves sprouting of capillaries followed by their invasion into the wounded area [111]. Angiogenesis is tightly controlled process. Cytokines and growth factors including PDGF, FGF-1, FGF-2, VEGF, TNF-α, MCP-1 and MIP-1α promote the angiogenesis [3, 112–114]. PDGF (especially PDGF-BB) and VEGF are most critical angiogenic growth factors. They mainly promote the formation of new vascular vessels by regulating the proliferation, migration and differentiation of endothelial cells [115, 116]. PDGF induces the production of other angiogenic factors, VEGF and FGF-2. VEGF is provided from wounded keratinocytes and macrophages in response to KGF and TGF-α. VEGF has an ability to promote the expression of endothelial specific NO-synthase (eNOS) expression. Nitric oxide (NO) is known as potent inducer of angiogenesis [117], and amplifies the VEGF-induced mitogenic response, indicating positive feedback system between NO and VEGF. Inhibitors of NO production antagonize the mitogenic effect of VEGF [118, 119]. VEGF is also induced in wound-edge keratinocytes and macrophages in response to KGF and TGF-α [120, 121]. In genetically diabetic mice, wound healing is impaired along with decreased VEGF expression at the wounded site [120]. Retroviral expression of dominant negative VEGF receptor-2 in murine skin wound results in severe defect in angiogenesis and granulation tissue formation, suggesting essential role of VEGF on wound angiogenesis and healing [122]. VEGF increases vascular permeability and induces fenestrations in the endothelium of small capillaries [123, 124]. This property of VEGF is important for angiogenesis in wound healing. FGF-2 and VEGF have synergistic effects on angiogenesis [125]. FGF-2 induces an angiogenic phenotype consisting of increased proliferation, migration, proteinase production, and expression of specific integrins [126, 127].

In response to the angiogenic signals, the endothelial cells in pre-existing vessels migrate into wounded site and sprout across the basement membrane. The expressions of $\alpha_v\beta_3$ and $\alpha_5\beta_1$ integrins are increased in migrating endothelial cells [128]. Blocking peptide or antibody against $\alpha_v\beta_3$ integrin inhibits the angiogenesis and wound healing [129]. Angiogenesis induced by FGF-2 or TNF-α is dependent on $\alpha_v\beta_3$ integrin, whereas VEGF or TNF-α-induced neovascularization is dependent on $\alpha_v\beta_5$ integrin [130]. Proteolysis of basement membrane surrounding existing blood vessels is required for migration of the endothelial cells through the basement membrane. At the activating front of growing vessels, activated endothelial cells release proteolytic enzyme, such as collagenase, to dissolve surrounding ECM [131]. After vascular tube formation, vascular wall is stabilized by incorporation of pericytes and smooth muscle cells. PDGF is involved in the recruitment of pericytes and smooth muscle cells.

Angiogenesis is inhibited by thrombospondin (TSP-1), IFN-γ, IL-4, IP-10 and tissue inhibitor of metalloproteinases (TIMPs). They inhibit the migration or proliferation of endothelial cells.

1.8 Remodeling

Early repair tissue is composed of a haphazard arrangement of collagen fibrils and abnormal component of proteoglycans and collagens [1, 2]. This lack of matrix order causes opacity and mechanical weakness of repair tissue. Breakdown and cross-linking of collagen fibril is required to stabilize the repaired tissue. As a result of tissue remodeling, granulation tissue changes into scar tissue, which is less cellular and less vascular than the granulation tissue. The blood vessels are refined to form a mature functional network [111]. Fibroblasts in the repair tissue continue to synthesis collagen. However, collagen production is reduced, even in the presence of TGF-β [1]. The decrease in cell density might result from migration of the cells out of the wound site or to apoptosis. The molecular mechanism of tissue remodeling is not fully understood. Remodeling of collagen fibril and other matrix proteins is driven by serine protease or MMPs secreted from fibroblasts under cytokines control. The coordinated regulation of protease and protease inhibitors may be important for tissue remodeling.

1.9 Resolution of Inflammation

Successful wound repair requires resolution of the inflammatory response. Incomplete resolution leads to scar formation and fibrosis [132, 133]. Fibrosis is an aberrant wound healing that caused by excess fibrous connective tissue. It is one of the main culprits for causing post-back or neck surgery problems. However, mechanisms that terminate immune response are poorly understood. In resolution phase, neutrophils are removed from a wounded site by apoptosis or by phagocytosis by macrophages [134]. Macrophages are inactivated by anti-inflammatory cytokine such as IL-10 [135]. IL-10 is a critical immunosuppressive cytokine that regulate both innate and adaptive immune response. Transplantation of skin graft from IL-10-deficient mice results in significantly higher infiltration of inflammatory cells [136]. Excess pro-inflammatory cytokines are removed by soluble cytokine receptors and receptor antagonists [137]. Expressions of anti-inflammatory molecules such as IL-1 receptor antagonist or soluble TNF receptor are increased in resolution phase. Secretory leukocyte protease inhibitor is a serine protease inhibitor, having anti-inflammatory activity. The deficiency of secretory leukocyto protease inhibitor enhances inflammatory response and delays wound healing [138].

There are many differences between fetal wound healing and adult wound healing. The embryonic tissue has the unique ability to heal wounds without scar formation and inflammation [139]. TGF-β is involved in multiple steps in wound healing, including recruitment of leukocytes, reepithelialization, fibroblast migration, wound contraction and tissue remodeling. The restricted scar formation in fetal wounds may be result from the different expression profiles of TGF-β isoforms. TGF-β3 is a predominant isoform in fetal wounds, though expressions of TGF-β1 and TGF-β2 are quite low [140, 141]. Furthermore, induction of TGF-β is restricted in fetal wound healing [142]. Application of exogenous TGF-β1 induces scar formation in fetal wounds [143, 144]. In adult wound healing, neutralizing antibody against TGF-β1 and TGF-β2 or exogenous application of TGF-β3 reduces scar formation [145]. These results suggest that TGF-β1 and TGF-β2 promotes the scar formation, whereas TGF-β3 has an opposite effect. Another feature of fetal wound healing is higher level of hyaluronan expression compared with adult granulation tissue [146]. The elevation of hyaluronan synthesis is associated with low levels of hyaluronidase. It has been shown that hyaluronan has inhibitory effects on scar formation [147]. It is possible that restricted scar formation in the fetal tissue may be result from the higher hyaluronan expression.

1.10   Proteases in Wound Healing

MMPs are endopeptidases that utilize a Zn^{2+} or Ca^{2+} ion in their active site. They play important roles in wound healing either directly degrade ECM components, or indirectly by their ability to regulate cell behavior. The major classes of MMPs involved in wound healing are collagenases (MMP-1, MMP-8), gelatinases (MMP-2, MMP-9), stromelysines (MMP-3, MMP-10), matrilysin (MMP-7), and membrane-type metalloproteases (MT-MMP; MMP-14) [148, 149]. Collagenases cleave type-I collagen, predominant collagen in dermis. Intact collagen fibrils are resistant to degradation by proteases other than collagenase. Gelatinases degrade type-I collagen to small fragments only after the initial cleavage of intact collagen by collagenases, and also have significant activity against type IV collagen, a major collagen in basement membrane. Stromelysins have broad substrate spectrum and are able to degrade ECM proteins including proteoglycans, fibronectin, laminin and non-fibrillar collagens (type IV and X). Matrilysin degrades proteoglycans, type IV collagen, fibronectin and elastin. MT-MMP is bound to cell membranes, and its activation plays a role in cell migration. Since MMPs are secreted into extracellular space as an inactive proenzyme. Another important biological role of MT-MMP is proteolytic activation of pro-MMP-2.

MMPs levels are increased in the early stage of wound healing [150, 151]. Macrophages and neutrophils secrete MMP-2, MMP-3 and MMP-9. They are involved in transmigration of leukocytes through blood vessel walls and infiltration into wounded area. MMP-8 released from neutrophils and macrophages plays important role in initial stage of wound healing by removing damaged ECM components at the wounded site.

Multiple MMPs are involved in reepithelialization. Treatment with MMP inhibitors attenuates keratinocyte migration and delays wound healing and reepithelialization [152, 153]. MMP-1 is required for keratinocyte-mediated reepithelialization as it cleaves components of cell-cell junctions and cell-ECM contacts in epithelium [154]. Cytokines and growth factor abundantly found in the wound environment promote MMPs expression. Upon injury, the expression of MMP-1 is induced in keratinocytes by EGF receptor-dependent mechanism [151, 154, 155]. TGF-β anti-sense oligonucleotide inhibits MMP-1 and MMP-9 production in keratinocytes [156]. Integrin $\alpha 2\beta 1$ upregulates MMP-1 and promotes migration of keratinocytes [157]. MMP-7 promotes reepithelialization by cleavage of E-cadherin within adherens junction. Overexpression of MMP-7 enhances keratinocyte migration, and loss of MMP-7 results in impaired wound healing [158]. Promoting effects of HGF and EGF on keratinocyte migration are dependent on MMP-9 activity [28]. MMP-10 expression is enhanced by KGF in migrating keratinocytes [159]. The activation of the MMPs is partly regulated by an extracellular calcium gradient, which is also required for the migration of the keratinocytes. Activation and relocalization of protein kinase C (PKC) to the cytoplasmic desmosomal plaque is responsible for decreasing adhesive properties of keratonocytes [160, 161]. Activated PKC upregulates MMP-2 activity in cultured human keratinocytes and stimulates proliferation, migration and wound closure [162].

MMP activity is also required for wound contraction. Impaired wound contraction is observed in mice-lacking MMP-3, whereas keratinocyte migration and reepithelialization is not affected by MMP-3 deficiency [163]. MMP expression in vascular endothelial cells is required for migration of vascular endothelial cells across the basement membrane during the formation of new capillaries. In the resolution phase of wound repair, MMPs secreted from fibroblasts are responsible for proteolytic degradation of ECM components.

The proteolytic activities of MMPs are inhibited by the tissue inhibitors of metallo-proteinases (TIMPs). TIMPs inhibit MMPs by interacting with active sites of MMPs. The existence of four kinds of TIMPs (TIMP-1, TIMP-2, TIMP-3, and TIMP-4) has been confirmed. TIMP-1 inhibits MMP-1, MMP-3 and MMP-9 more effectively than TIMP-2. TIMP-2 inhibits proMMP-2 over tenfold more effectively than TIMP-1. TIMP-3 inhibits at least MMP-2 and MMP-9. Interestingly, overexpression of MMP-1 or TIMP-1 results in impaired wound healing [164, 165], suggesting that the balance between MMPs and TIMPs may be a critical for precise regulation of cell migration and ECM deposition during wound healing.

Fibrin disrupts adhesion of differentiated keratinocytes and promotes keratinocyte migration indirectly by exposing plasminogen to migrating cells [166]. Plasmin is a fibrinolytic enzyme, derived from plasminogen. The urokinase plasminogen activator (uPA) system regulates keratinocyte migration through proteolysis during epithelial repair. uPA secreted from keratinocytes binds to cell surface by a specific receptor (uPA-R). uPA activates plasminogen to form plasmin, which participates in tissue degradation and proteolysis [167]. In addition, the uPA/uPAR complex can activate cellular responses independent of the proteolytic activity.

Plasminogen activator inhibitor type-1 (PAI-1) is also involved in two different processes during epithelial wound healing [168, 169]. PAI-1 acts as an inhibitor of the proteolytic pathway by preventing plasminogen activation by uPA. On the other hand, PAI-1 acts as a part of transitory anchoring the uPA/PAR complex. The interactions of uPA/uPAR/PAI-1 complex, with transmembrane receptors lead to the activation of intracellular signaling machinery like MAP kinases. The uPAR/uPA complex activates signal transduction pathway by low-density lipoprotein receptor-related protein (LRP) dependent mechanism, and promotes the motility of keratinocytes.

1.11   Anti-microbial Peptides

Microbial components impaired wound healing by inducing persisting inflammatory response. Keratinocytes play an important role in the innate immune response [170]. They produce anti-microbial peptides such as β-defensin-1, β-defensin-2, β-defensin-3 and CAP-18. CAP-18 is a member of cathelicidin family, and processed into LL-37 peptide by proteolytic cleavage. Release of β-defensin-2 is dependent on IL-1 activation by Toll-like receptors (TLR). CAP-18 release is mediated by trans-activation of EGFR by HB-EGF [171]. They were released in response to injury or exposure of microbial components. The release of β-defensin-2 and CAP-18 is transiently increased in wounded human skin. In addition, β-defensin-2 promotes the migration of keratinocytes, and subsequent reepithelialization. Further study is required to elucidate the physiological role of the antimicrobial peptides.

1.12   Chronic Wound Healing

As described, wound healing is a complex process. Cellular functions involved in acute wound healing are tightly regulated by numerous growth factors, cytokines and proteases. Thus, spatial and temporal alterations in the actions of these molecules result in the failure of wound healing. In many case, wound healing disorders lead to non-healing chronic wounds. Incidence of chronic wounds is higher among aged or diabetic patients. A special feature of chronic wound is a prolonged or excessive inflammatory response at the wounded site [172]. It is characterized by excess infiltration of neutrophils and macrophages. They provide pro-inflammatory cytokines (IL-1, IL-6 and TNF-α), ROS (hydrogen peroxide, superoxide anion, hydroxy radicals and singlet oxygen) and proteases (MMP-8, leukocyto elastase) at the wound site. Pro-inflammatory cytokines are a powerful inducer of MMPs (MMP-1, MMP-3, MMP-9 and MMP-13) while down regulating TIMPs expression at the chronic wound site. The elevated MMP activity and down regulation of TIMPs leads to degradation of the extracellular matrix and growth factors [173], which further inhibits the wound healing process. MMPs released from neutrophils degrade important healing factors such as PDGF and TGF-β. Impaired wound healing in the aged mice is associated with a delay in PDGF and PDGF receptor expression [174]. In non-healing dermal ulcers, the levels of PDGF are reduced [175].

Excess ROS, provided from neutrophils and macrophages, directly damages cell membrane and structural proteins of ECM. Furthermore, ROS amplifies the persistent inflammatory state of chronic wounds by activating transcription factors that control the pro-inflammatory cytokines and proteases [176]. Biochemical analyses of fluid and biopsies from healing and chronic wound indicate that the molecular environment of healing wounds is high levels of growth factors and cytokines that promote cell migration, low levels of inflammatory cytokines and proteases. In contrast, molecular environment of chronic wound is reduced mitogenic activity for fibroblasts and keratinocytes [177]. The cytokine environment of chronic wound is more pro-inflammatory than acute wounds. In chronic wounds, the levels of IL-1, IL-6 and TNF-α are greatly higher than healing wounds. The epidermis of chronic wound is thick and hyper-proliferative and hyperkeratotic appearance. Keratinocytes of chronic wound edge are capable of proliferation but are unable to migrate into a wounded site. Persistent activation of c-Myc in basal keratinocyte is correlated with inhibition of keratinocyte migration.

In an attempt to heal chronic wounds, different strategies are introduced, such as administration of growth factors or hyaluronan, transplantation of fibroblasts or epithelial stem cells. These treatments are only partially successful. It should be focused on the prevention of excess inflammatory response which results from infiltration of neutrophils and macrophage at a wound site.

References

1. Martin P (1997) Wound healing–aiming for perfect skin regeneration. Science 276(5309):75–81
2. Shaw TJ, Martin P (2009) Wound repair at a glance. J Cell Sci 122(Pt 18):3209–3213
3. Werner S, Grose R (2003) Regulation of wound healing by growth factors and cytokines. Physiol Rev 83(3):835–870
4. Noli C, Miolo A (2001) The mast cell in wound healing. Vet Dermatol 12(6):303–313
5. Jameson JM, Sharp LL, Witherden DA, Havran WL (2004) Regulation of skin cell homeostasis by gamma delta T cells. Front Biosci 9:2640–2651
6. Cumberbatch M, Dearman RJ, Griffiths CE, Kimber I (2000) Langerhans cell migration. Clin Exp Dermatol 25(5):413–418

7. Engelhardt E, Toksoy A, Goebeler M, Debus S et al (1998) Chemokines IL-8, GROalpha, MCP-1, IP-10, and Mig are sequentially and differentially expressed during phase-specific infiltration of leukocyte subsets in human wound healing. Am J Pathol 153(6):1849–1860

8. Jackman SH, Yoak MB, Keerthy S, Beaver BL (2000) Differential expression of chemokines in a mouse model of wound healing. Ann Clin Lab Sci 30(2):201–207

9. Gibran NS, Ferguson M, Heimbach DM, Isik FF (1997) Monocyte chemoattractant protein-1 mRNA expression in the human burn wound. J Surg Res 70(1):1–6

10. Low QE, Drugea IA, Duffner LA, Quinn DG et al (2001) Wound healing in MIP-1alpha(−/−) and MCP-1(−/−) mice. Am J Pathol 159(2):457–463

11. Nagaoka T, Kaburagi Y, Hamaguchi Y, Hasegawa M et al (2000) Delayed wound healing in the absence of intercellular adhesion molecule-1 or L-selectin expression. Am J Pathol 157(1):237–247

12. Jung U, Ley K (1999) Mice lacking two or all three selectins demonstrate overlapping and distinct functions for each selectin. J Immunol 162(11):6755–6762

13. Hubner G, Brauchle M, Smola H, Madlener M et al (1996) Differential regulation of pro-inflammatory cytokines during wound healing in normal and glucocorticoid-treated mice. Cytokine 8(7):548–556

14. Mori R, Shaw TJ, Martin P (2008) Molecular mechanisms linking wound inflammation and fibrosis: knockdown of osteopontin leads to rapid repair and reduced scarring. J Exp Med 205(1):43–51

15. Martinez FO, Sica A, Mantovani A, Locati M (2008) Macrophage activation and polarization. Front Biosci 13:453–461

16. Eming SA, Krieg T, Davidson JM (2007) Inflammation in wound repair: molecular and cellular mechanisms. J Invest Dermatol 127(3):514–525

17. DiPietro LA, Polverini PJ (1993) Role of the macrophage in the positive and negative regulation of wound neovascularization. Behring Inst Mitt 92:238–247

18. Leibovich SJ, Ross R (1975) The role of the macrophage in wound repair. A study with hydrocortisone and antimacrophage serum. Am J Pathol 78(1):71–100

19. Egozi EI, Ferreira AM, Burns AL, Gamelli RL et al (2003) Mast cells modulate the inflammatory but not the proliferative response in healing wounds. Wound Repair Regen 11(1):46–54

20. Jameson J, Ugarte K, Chen N, Yachi P et al (2002) A role for skin gammadelta T cells in wound repair. Science 296(5568):747–749

21. Jameson JM, Cauvi G, Sharp LL, Witherden DA et al (2005) Gammadelta T cell-induced hyaluronan production by epithelial cells regulates inflammation. J Exp Med 201(8):1269–1279

22. Singer AJ, Clark RA (1999) Cutaneous wound healing. N Engl J Med 341(10):738–746

23. Lee JM, Dedhar S, Kalluri R, Thompson EW (2006) The epithelial-mesenchymal transition: new insights in signaling, development, and disease. J Cell Biol 172(7):973–981

24. Feldmeyer L, Werner S, French LE, Beer HD (2010) Interleukin-1, inflammasomes and the skin. Eur J Cell Biol 89(9):638–644

25. Higashiyama M, Matsumoto K, Hashimoto K, Yoshikawa K (1991) Increased production of transforming growth factor-alpha in psoriatic epidermis. J Dermatol 18(2):117–119

26. Barrandon Y, Green H (1987) Cell migration is essential for sustained growth of keratinocyte colonies: the roles of transforming growth factor-alpha and epidermal growth factor. Cell 50(7):1131–1137

27. Werner S, Breeden M, Hubner G, Greenhalgh DG et al (1994) Induction of keratinocyte growth factor expression is reduced and delayed during wound healing in the genetically diabetic mouse. J Invest Dermatol 103(4):469–473

28. McCawley LJ, O'Brien P, Hudson LG (1998) Epidermal growth factor (EGF)- and scatter factor/hepatocyte growth factor (SF/HGF)- mediated keratinocyte migration is coincident with induction of matrix metalloproteinase (MMP)-9. J Cell Physiol 176(2):255–265

29. Raja SK, Garcia MS, Isseroff RR (2007) Wound re-epithelialization: modulating keratinocyte migration in wound healing. Front Biosci 12:2849–2868

30. Hebda PA (1988) Stimulatory effects of transforming growth factor-beta and epidermal growth factor on epidermal cell outgrowth from porcine skin explant cultures. J Invest Dermatol 91(5):440–445

31. Li Y, Fan J, Chen M, Li W et al (2006) Transforming growth factor-alpha: a major human serum factor that promotes human keratinocyte migration. J Invest Dermatol 126(9):2096–2105
32. Hashimoto K, Higashiyama S, Asada H, Hashimura E et al (1994) Heparin-binding epidermal growth factor-like growth factor is an autocrine growth factor for human keratinocytes. J Biol Chem 269(31):20060–20066
33. Brown GL, Curtsinger L 3rd, Brightwell JR, Ackerman DM et al (1986) Enhancement of epidermal regeneration by biosynthetic epidermal growth factor. J Exp Med 163(5):1319–1324
34. Schultz GS, White M, Mitchell R, Brown G et al (1987) Epithelial wound healing enhanced by transforming growth factor-alpha and vaccinia growth factor. Science 235(4786):350–352
35. Kim I, Mogford JE, Chao JD, Mustoe TA (2001) Wound epithelialization deficits in the transforming growth factor-alpha knockout mouse. Wound Repair Regen 9(5):386–390
36. Shirakata Y, Kimura R, Nanba D, Iwamoto R et al (2005) Heparin-binding EGF-like growth factor accelerates keratinocyte migration and skin wound healing. J Cell Sci 118(Pt 11):2363–2370
37. Bhora FY, Dunkin BJ, Batzri S, Aly HM et al (1995) Effect of growth factors on cell proliferation and epithelialization in human skin. J Surg Res 59(2):236–244
38. DeLapp NW, Dieckman DK (1990) Effect of basic fibroblast growth factor (bFGF) and insulin-like growth factors type I (IGF-I) and type II (IGF-II) on adult human keratinocyte growth and fibronectin secretion. J Invest Dermatol 94(6):777–780
39. Hebda PA, Klingbeil CK, Abraham JA, Fiddes JC (1990) Basic fibroblast growth factor stimulation of epidermal wound healing in pigs. J Invest Dermatol 95(6):626–631
40. Sanz Garcia S, Santos Heredero X, Izquierdo Hernandez A, Pascual Pena E et al (2000) Experimental model for local application of growth factors in skin re-epithelialisation. Scand J Plast Reconstr Surg Hand Surg 34(3):199–206
41. Ortega S, Ittmann M, Tsang SH, Ehrlich M et al (1998) Neuronal defects and delayed wound healing in mice lacking fibroblast growth factor 2. Proc Natl Acad Sci USA 95(10):5672–5677
42. Miller DL, Ortega S, Bashayan O, Basch R et al (2000) Compensation by fibroblast growth factor 1 (FGF1) does not account for the mild phenotypic defects observed in FGF2 null mice. Mol Cell Biol 20(6):2260–2268
43. Werner S, Peters KG, Longaker MT, Fuller-Pace F et al (1992) Large induction of keratinocyte growth factor expression in the dermis during wound healing. Proc Natl Acad Sci USA 89(15):6896–6900
44. Brauchle M, Fassler R, Werner S (1995) Suppression of keratinocyte growth factor expression by glucocorticoids in vitro and during wound healing. J Invest Dermatol 105(4):579–584
45. Guo L, Degenstein L, Fuchs E (1996) Keratinocyte growth factor is required for hair development but not for wound healing. Genes Dev 10(2):165–175
46. Ashcroft GS (1999) Bidirectional regulation of macrophage function by TGF-beta. Microbes Infect 1(15):1275–1282
47. Huang JS, Wang YH, Ling TY, Chuang SS et al (2002) Synthetic TGF-beta antagonist accelerates wound healing and reduces scarring. FASEB J 16(10):1269–1270
48. Tredget EB, Demare J, Chandran G, Tredget EE et al (2005) Transforming growth factor-beta and its effect on reepithelialization of partial-thickness ear wounds in transgenic mice. Wound Repair Regen 13(1):61–67
49. Sharma GD, He J, Bazan HE (2003) p38 and ERK1/2 coordinate cellular migration and proliferation in epithelial wound healing: evidence of cross-talk activation between MAP kinase cascades. J Biol Chem 278(24):21989–21997
50. Tscharntke M, Pofahl R, Chrostek-Grashoff A, Smyth N et al (2007) Impaired epidermal wound healing in vivo upon inhibition or deletion of Rac1. J Cell Sci 120(Pt 8):1480–1490
51. Tscharntke M, Pofahl R, Krieg T, Haase I (2005) Ras-induced spreading and wound closure in human epidermal keratinocytes. FASEB J 19(13):1836–1838
52. Pullar CE, Baier BS, Kariya Y, Russell AJ et al (2006) beta4 integrin and epidermal growth factor coordinately regulate electric field-mediated directional migration via Rac1. Mol Biol Cell 17(11):4925–4935

53. Haase I, Evans R, Pofahl R, Watt FM (2003) Regulation of keratinocyte shape, migration and wound epithelialization by IGF-1- and EGF-dependent signalling pathways. J Cell Sci 116(Pt 15):3227–3238

54. Schweizer J, Bowden PE, Coulombe PA, Langbein L et al (2006) New consensus nomenclature for mammalian keratins. J Cell Biol 174(2):169–174

55. Guo L, Degenstein L, Dowling J, Yu QC et al (1995) Gene targeting of BPAG1: abnormalities in mechanical strength and cell migration in stratified epithelia and neurologic degeneration. Cell 81(2):233–243

56. Stoler A, Kopan R, Duvic M, Fuchs E (1988) Use of monospecific antisera and cRNA probes to localize the major changes in keratin expression during normal and abnormal epidermal differentiation. J Cell Biol 107(2):427–446

57. Mansbridge JN, Knapp AM (1987) Changes in keratinocyte maturation during wound healing. J Invest Dermatol 89(3):253–263

58. Jiang CK, Magnaldo T, Ohtsuki M, Freedberg IM et al (1993) Epidermal growth factor and transforming growth factor alpha specifically induce the activation- and hyperproliferation-associated keratins 6 and 16. Proc Natl Acad Sci USA 90(14):6786–6790

59. Komine M, Rao LS, Kaneko T, Tomic-Canic M et al (2000) Inflammatory versus proliferative processes in epidermis. Tumor necrosis factor alpha induces K6b keratin synthesis through a transcriptional complex containing NFkappa B and C/EBPbeta. J Biol Chem 275(41):32077–32088

60. Komine M, Rao LS, Freedberg IM, Simon M et al (2001) Interleukin-1 induces transcription of keratin K6 in human epidermal keratinocytes. J Invest Dermatol 116(2):330–338

61. Wawersik M, Coulombe PA (2000) Forced expression of keratin 16 alters the adhesion, differentiation, and migration of mouse skin keratinocytes. Mol Biol Cell 11(10):3315–3327

62. Wong P, Coulombe PA (2003) Loss of keratin 6 (K6) proteins reveals a function for intermediate filaments during wound repair. J Cell Biol 163(2):327–337

63. Margadant C, Charafeddine RA, Sonnenberg A (2010) Unique and redundant functions of integrins in the epidermis. FASEB J 24(11):4133–4152

64. Borradori L, Sonnenberg A (1999) Structure and function of hemidesmosomes: more than simple adhesion complexes. J Invest Dermatol 112(4):411–418

65. Rabinovitz I, Tsomo L, Mercurio AM (2004) Protein kinase C-alpha phosphorylation of specific serines in the connecting segment of the beta 4 integrin regulates the dynamics of type II hemidesmosomes. Mol Cell Biol 24(10):4351–4360

66. Dans M, Gagnoux-Palacios L, Blaikie P, Klein S et al (2001) Tyrosine phosphorylation of the beta 4 integrin cytoplasmic domain mediates Shc signaling to extracellular signal-regulated kinase and antagonizes formation of hemidesmosomes. J Biol Chem 276(2):1494–1502

67. Wilhelmsen K, Litjens SH, Sonnenberg A (2006) Multiple functions of the integrin alpha-6beta4 in epidermal homeostasis and tumorigenesis. Mol Cell Biol 26(8):2877–2886

68. Wilhelmsen K, Litjens SH, Kuikman I, Margadant C et al (2007) Serine phosphorylation of the integrin beta4 subunit is necessary for epidermal growth factor receptor induced hemidesmosome disruption. Mol Biol Cell 18(9):3512–3522

69. Larjava H, Salo T, Haapasalmi K, Kramer RH et al (1993) Expression of integrins and basement membrane components by wound keratinocytes. J Clin Invest 92(3):1425–1435

70. Adams JC, Watt FM (1991) Expression of beta 1, beta 3, beta 4, and beta 5 integrins by human epidermal keratinocytes and non-differentiating keratinocytes. J Cell Biol 115(3):829–841

71. Cavani A, Zambruno G, Marconi A, Manca V et al (1993) Distinctive integrin expression in the newly forming epidermis during wound healing in humans. J Invest Dermatol 101(4):600–604

72. Breuss JM, Gallo J, DeLisser HM, Klimanskaya IV et al (1995) Expression of the beta 6 integrin subunit in development, neoplasia and tissue repair suggests a role in epithelial remodeling. J Cell Sci 108(Pt 6):2241–2251

73. Haapasalmi K, Zhang K, Tonnesen M, Olerud J et al (1996) Keratinocytes in human wounds express alpha v beta 6 integrin. J Invest Dermatol 106(1):42–48

74. Grose R, Hutter C, Bloch W, Thorey I et al (2002) A crucial role of beta 1 integrins for keratinocyte migration in vitro and during cutaneous wound repair. Development 129(9): 2303–2315

75. Calvin M (1998) Cutaneous wound repair. Wounds Compend Clin Res Pract 10(1):12–32

76. Frank S, Madlener M, Werner S (1996) Transforming growth factors beta1, beta2, and beta3 and their receptors are differentially regulated during normal and impaired wound healing. J Biol Chem 271(17):10188–10193

77. Buetow BS, Crosby JR, Kaminski WE, Ramachandran RK et al (2001) Platelet-derived growth factor B-chain of hematopoietic origin is not necessary for granulation tissue formation and its absence enhances vascularization. Am J Pathol 159(5):1869–1876

78. Heldin CH, Westermark B (1999) Mechanism of action and in vivo role of platelet-derived growth factor. Physiol Rev 79(4):1283–1316

79. Roberts AB, Sporn MB, Assoian RK, Smith JM et al (1986) Transforming growth factor type beta: rapid induction of fibrosis and angiogenesis in vivo and stimulation of collagen formation in vitro. Proc Natl Acad Sci USA 83(12):4167–4171

80. Brown RL, Ormsby I, Doetschman TC, Greenhalgh DG (1995) Wound healing in the transforming growth factor-beta-deficient mouse. Wound Repair Regen 3(1):25–36

81. Bonomo SR, Davidson JD, Tyrone JW, Lin X et al (2000) Enhancement of wound healing by hyperbaric oxygen and transforming growth factor beta3 in a new chronic wound model in aged rabbits. Arch Surg 135(10):1148–1153

82. Bradham DM, Igarashi A, Potter RL, Grotendorst GR (1991) Connective tissue growth factor: a cysteine-rich mitogen secreted by human vascular endothelial cells is related to the SRC-induced immediate early gene product CEF-10. J Cell Biol 114(6):1285–1294

83. Clark RA (1990) Fibronectin matrix deposition and fibronectin receptor expression in healing and normal skin. J Invest Dermatol 94(6 Suppl):128S–134S

84. Xu J, Clark RA (1996) Extracellular matrix alters PDGF regulation of fibroblast integrins. J Cell Biol 132(1–2):239–249

85. Grinnell F (1994) Fibroblasts, myofibroblasts, and wound contraction. J Cell Biol 124(4):401–404

86. Coleman C, Tuan TL, Buckley S, Anderson KD et al (1998) Contractility, transforming growth factor-beta, and plasmin in fetal skin fibroblasts: role in scarless wound healing. Pediatr Res 43(3):403–409

87. Nedelec B, Ghahary A, Scott PG, Tredget EE (2000) Control of wound contraction. Basic and clinical features. Hand Clin 16(2):289–302

88. Grinnell F, Ho CH, Lin YC, Skuta G (1999) Differences in the regulation of fibroblast contraction of floating versus stressed collagen matrices. J Biol Chem 274(2):918–923

89. Tingstrom A, Heldin CH, Rubin K (1992) Regulation of fibroblast-mediated collagen gel contraction by platelet-derived growth factor, interleukin-1 alpha and transforming growth factor-beta 1. J Cell Sci 102(Pt 2):315–322

90. Moulin V, Castilloux G, Auger FA, Garrel D et al (1998) Modulated response to cytokines of human wound healing myofibroblasts compared to dermal fibroblasts. Exp Cell Res 238(1):283–293

91. Lee YR, Oshita Y, Tsuboi R, Ogawa H (1996) Combination of insulin-like growth factor (IGF)-I and IGF-binding protein-1 promotes fibroblast-embedded collagen gel contraction. Endocrinology 137(12):5278–5283

92. Pilcher BK, Levine NS, Tomasek JJ (1995) Thrombin promotion of isometric contraction in fibroblasts: its extracellular mechanism of action. Plast Reconstr Surg 96(5): 1188–1195

93. Gullberg D, Tingstrom A, Thuresson AC, Olsson L et al (1990) Beta 1 integrin-mediated collagen gel contraction is stimulated by PDGF. Exp Cell Res 186(2):264–272

94. Carver W, Molano I, Reaves TA, Borg TK et al (1995) Role of the alpha 1 beta 1 integrin complex in collagen gel contraction in vitro by fibroblasts. J Cell Physiol 165(2):425–437

95. Langholz O, Rockel D, Mauch C, Kozlowska E et al (1995) Collagen and collagenase gene expression in three-dimensional collagen lattices are differentially regulated by alpha 1 beta 1 and alpha 2 beta 1 integrins. J Cell Biol 131(6 Pt 2):1903–1915

96. Riikonen T, Westermarck J, Koivisto L, Broberg A et al (1995) Integrin alpha 2 beta 1 is a positive regulator of collagenase (MMP-1) and collagen alpha 1(I) gene expression. J Biol Chem 270(22):13548–13552

97. Iocono JA, Colleran KR, Remick DG, Gillespie BW et al (2000) Interleukin-8 levels and activity in delayed-healing human thermal wounds. Wound Repair Regen 8(3):216–225

98. Zhu YK, Liu XD, Skold MC, Umino T et al (2001) Cytokine inhibition of fibroblast-induced gel contraction is mediated by PGE(2) and NO acting through separate parallel pathways. Am J Respir Cell Mol Biol 25(2):245–253

99. Pellegrin S, Mellor H (2007) Actin stress fibres. J Cell Sci 120(Pt 20):3491–3499

100. Paterson HF, Self AJ, Garrett MD, Just I et al (1990) Microinjection of recombinant p21rho induces rapid changes in cell morphology. J Cell Biol 111(3):1001–1007

101. Katoh K, Kano Y, Amano M, Onishi H et al (2001) Rho-kinase–mediated contraction of isolated stress fibers. J Cell Biol 153(3):569–584

102. Somlyo AP, Somlyo AV (2000) Signal transduction by G-proteins, rho-kinase and protein phosphatase to smooth muscle and non-muscle myosin II. J Physiol 522(Pt 2):177–185

103. Amano M, Ito M, Kimura K, Fukata Y et al (1996) Phosphorylation and activation of myosin by Rho-associated kinase (Rho-kinase). J Biol Chem 271(34):20246–20249

104. Kimura K, Ito M, Amano M, Chihara K et al (1996) Regulation of myosin phosphatase by Rho and Rho-associated kinase (Rho-kinase). Science 273(5272):245–248

105. Koyama M, Ito M, Feng J, Seko T et al (2000) Phosphorylation of CPI-17, an inhibitory phosphoprotein of smooth muscle myosin phosphatase, by Rho-kinase. FEBS Lett 475(3):197–200

106. Kitazawa T, Eto M, Woodsome TP, Brautigan DL (2000) Agonists trigger G protein-mediated activation of the CPI-17 inhibitor phosphoprotein of myosin light chain phosphatase to enhance vascular smooth muscle contractility. J Biol Chem 275(14):9897–9900

107. Katoh K, Kano Y, Noda Y (2011) Rho-associated kinase-dependent contraction of stress fibres and the organization of focal adhesions. J R Soc Interface 8(56):305–311

108. Hinz B (2007) Formation and function of the myofibroblast during tissue repair. J Invest Dermatol 127(3):526–537

109. Desmouliere A, Geinoz A, Gabbiani F, Gabbiani G (1993) Transforming growth factor-beta 1 induces alpha-smooth muscle actin expression in granulation tissue myofibroblasts and in quiescent and growing cultured fibroblasts. J Cell Biol 122(1):103–111

110. Hinz B, Mastrangelo D, Iselin CE, Chaponnier C et al (2001) Mechanical tension controls granulation tissue contractile activity and myofibroblast differentiation. Am J Pathol 159(3):1009–1020

111. Adams RH, Alitalo K (2007) Molecular regulation of angiogenesis and lymphangiogenesis. Nat Rev Mol Cell Biol 8(6):464–478

112. Liekens S, De Clercq E, Neyts J (2001) Angiogenesis: regulators and clinical applications. Biochem Pharmacol 61(3):253–270

113. Conway EM, Collen D, Carmeliet P (2001) Molecular mechanisms of blood vessel growth. Cardiovasc Res 49(3):507–521

114. Munoz-Chapuli R, Quesada AR, Angel Medina M (2004) Angiogenesis and signal transduction in endothelial cells. Cell Mol Life Sci 61(17):2224–2243

115. Battegay EJ, Rupp J, Iruela-Arispe L, Sage EH et al (1994) PDGF-BB modulates endothelial proliferation and angiogenesis in vitro via PDGF beta-receptors. J Cell Biol 125(4):917–928

116. Ferrara N (1999) Role of vascular endothelial growth factor in the regulation of angiogenesis. Kidney Int 56(3):794–814

117. Cooke JP, Losordo DW (2002) Nitric oxide and angiogenesis. Circulation 105(18):2133–2135

118. Morbidelli L, Chang CH, Douglas JG, Granger HJ et al (1996) Nitric oxide mediates mitogenic effect of VEGF on coronary venular endothelium. Am J Physiol 270(1 Pt 2):H411–H415

119. Parenti A, Morbidelli L, Cui XL, Douglas JG et al (1998) Nitric oxide is an upstream signal of vascular endothelial growth factor-induced extracellular signal-regulated kinase1/2 activation in postcapillary endothelium. J Biol Chem 273(7):4220–4226

120. Frank S, Hubner G, Breier G, Longaker MT et al (1995) Regulation of vascular endothelial growth factor expression in cultured keratinocytes. Implications for normal and impaired wound healing. J Biol Chem 270(21):12607–12613

121. Brown LF, Yeo KT, Berse B, Yeo TK et al (1992) Expression of vascular permeability factor (vascular endothelial growth factor) by epidermal keratinocytes during wound healing. J Exp Med 176(5):1375–1379

122. Tsou R, Fathke C, Wilson L, Wallace K et al (2002) Retroviral delivery of dominant-negative vascular endothelial growth factor receptor type 2 to murine wounds inhibits wound angiogenesis. Wound Repair Regen 10(4):222–229

123. Senger DR, Galli SJ, Dvorak AM, Perruzzi CA et al (1983) Tumor cells secrete a vascular permeability factor that promotes accumulation of ascites fluid. Science 219(4587):983–985

124. Roberts WG, Palade GE (1995) Increased microvascular permeability and endothelial fenestration induced by vascular endothelial growth factor. J Cell Sci 108(Pt 6):2369–2379

125. Pepper MS, Ferrara N, Orci L, Montesano R (1992) Potent synergism between vascular endothelial growth factor and basic fibroblast growth factor in the induction of angiogenesis in vitro. Biochem Biophys Res Commun 189(2):824–831

126. Klein S, Giancotti FG, Presta M, Albelda SM et al (1993) Basic fibroblast growth factor modulates integrin expression in microvascular endothelial cells. Mol Biol Cell 4(10):973–982

127. Klein S, Bikfalvi A, Birkenmeier TM, Giancotti FG et al (1996) Integrin regulation by endogenous expression of 18-kDa fibroblast growth factor-2. J Biol Chem 271(37):22583–22590

128. Ruegg C, Mariotti A (2003) Vascular integrins: pleiotropic adhesion and signaling molecules in vascular homeostasis and angiogenesis. Cell Mol Life Sci 60(6):1135–1157

129. Brooks PC, Clark RA, Cheresh DA (1994) Requirement of vascular integrin alpha v beta 3 for angiogenesis. Science 264(5158):569–571

130. Friedlander M, Brooks PC, Shaffer RW, Kincaid CM et al (1995) Definition of two angiogenic pathways by distinct alpha v integrins. Science 270(5241):1500–1502

131. Fisher C, Gilbertson-Beadling S, Powers EA, Petzold G et al (1994) Interstitial collagenase is required for angiogenesis in vitro. Dev Biol 162(2):499–510

132. Coussens LM, Werb Z (2002) Inflammation and cancer. Nature 420(6917):860–867

133. Wynn TA (2008) Cellular and molecular mechanisms of fibrosis. J Pathol 214(2):199–210

134. Haslett C (1992) Resolution of acute inflammation and the role of apoptosis in the tissue fate of granulocytes. Clin Sci (Lond) 83(6):639–648

135. Ma J, Chen T, Mandelin J, Ceponis A et al (2003) Regulation of macrophage activation. Cell Mol Life Sci 60(11):2334–2346

136. Liechty KW, Kim HB, Adzick NS and Crombleholme TM (2000) Fetal wound repair results in scar formation in interleukin-10-deficient mice in a syngeneic murine model of scarless fetal wound repair. J Pediatr Surg 35(6), 866–872; discussion 872–863

137. D'Amico G, Frascaroli G, Bianchi G, Transidico P et al (2000) Uncoupling of inflammatory chemokine receptors by IL-10: generation of functional decoys. Nat Immunol 1(5):387–391

138. Ashcroft GS, Lei K, Jin W, Longenecker G et al (2000) Secretory leukocyte protease inhibitor mediates non-redundant functions necessary for normal wound healing. Nat Med 6(10): 1147–1153

139. McCluskey J, Martin P (1995) Analysis of the tissue movements of embryonic wound healing–DiI studies in the limb bud stage mouse embryo. Dev Biol 170(1):102–114

140. Sullivan KM, Lorenz HP, Meuli M, Lin RY et al. (1995) A model of scarless human fetal wound repair is deficient in transforming growth factor beta. J Pediatr Surg 30(2), 198–202; discussion 202–193

141. Hsu M, Peled ZM, Chin GS, Liu W et al. (2001) Ontogeny of expression of transforming growth factor-beta 1 (TGF-beta 1), TGF-beta 3, and TGF-beta receptors I and II in fetal rat fibroblasts and skin. Plast Reconstr Surg 107(7), 1787–1794; discussion 1795–1786

142. Nath RK, LaRegina M, Markham H, Ksander GA et al (1994) The expression of transforming growth factor type beta in fetal and adult rabbit skin wounds. J Pediatr Surg 29(3):416–421

143. Krummel TM, Michna BA, Thomas BL, Sporn MB et al (1988) Transforming growth factor beta (TGF-beta) induces fibrosis in a fetal wound model. J Pediatr Surg 23(7):647–652

144. Lin RY, Sullivan KM, Argenta PA, Meuli M et al (1995) Exogenous transforming growth factor-beta amplifies its own expression and induces scar formation in a model of human fetal skin repair. Ann Surg 222(2):146–154

145. Shah M, Foreman DM, Ferguson MW (1995) Neutralisation of TGF-beta 1 and TGF-beta 2 or exogenous addition of TGF-beta 3 to cutaneous rat wounds reduces scarring. J Cell Sci 108(Pt 3):985–1002

146. Longaker MT, Chiu ES, Adzick NS, Stern M et al (1991) Studies in fetal wound healing. V. A prolonged presence of hyaluronic acid characterizes fetal wound fluid. Ann Surg 213(4):292–296

147. Hellstrom S, Laurent C (1987) Hyaluronan and healing of tympanic membrane perforations. An experimental study. Acta Otolaryngol Suppl 442:54–61

148. Gill SE, Parks WC (2008) Metalloproteinases and their inhibitors: regulators of wound healing. Int J Biochem Cell Biol 40(6–7):1334–1347

149. Toriseva M, Kahari VM (2009) Proteinases in cutaneous wound healing. Cell Mol Life Sci 66(2):203–224

150. Salo T, Makela M, Kylmaniemi M, Autio-Harmainen H et al (1994) Expression of matrix metalloproteinase-2 and -9 during early human wound healing. Lab Invest 70(2):176–182

151. Sudbeck BD, Pilcher BK, Welgus HG, Parks WC (1997) Induction and repression of collagenase-1 by keratinocytes is controlled by distinct components of different extracellular matrix compartments. J Biol Chem 272(35):22103–22110

152. Agren MS (1999) Matrix metalloproteinases (MMPs) are required for re-epithelialization of cutaneous wounds. Arch Dermatol Res 291(11):583–590

153. Mirastschijski U, Haaksma CJ, Tomasek JJ, Agren MS (2004) Matrix metalloproteinase inhibitor GM 6001 attenuates keratinocyte migration, contraction and myofibroblast formation in skin wounds. Exp Cell Res 299(2):465–475

154. Pilcher BK, Dumin JA, Sudbeck BD, Krane SM et al (1997) The activity of collagenase-1 is required for keratinocyte migration on a type I collagen matrix. J Cell Biol 137(6):1445–1457

155. Pilcher BK, Dumin J, Schwartz MJ, Mast BA et al (1999) Keratinocyte collagenase-1 expression requires an epidermal growth factor receptor autocrine mechanism. J Biol Chem 274(15):10372–10381

156. Philipp K, Riedel F, Germann G, Hormann K et al (2005) TGF-beta antisense oligonucleotides reduce mRNA expression of matrix metalloproteinases in cultured wound-healing-related cells. Int J Mol Med 15(2):299–303

157. Pilcher BK, Sudbeck BD, Dumin JA, Welgus HG et al (1998) Collagenase-1 and collagen in epidermal repair. Arch Dermatol Res 290(Suppl):S37–S46

158. McGuire JK, Li Q, Parks WC (2003) Matrilysin (matrix metalloproteinase-7) mediates E-cadherin ectodomain shedding in injured lung epithelium. Am J Pathol 162(6):1831–1843

159. Tsuboi R, Sato C, Kurita Y, Ron D et al (1993) Keratinocyte growth factor (FGF-7) stimulates migration and plasminogen activator activity of normal human keratinocytes. J Invest Dermatol 101(1):49–53

160. Garrod DR, Berika MY, Bardsley WF, Holmes D et al (2005) Hyper-adhesion in desmosomes: its regulation in wound healing and possible relationship to cadherin crystal structure. J Cell Sci 118(Pt 24):5743–5754

161. Wallis S, Lloyd S, Wise I, Ireland G et al (2000) The alpha isoform of protein kinase C is involved in signaling the response of desmosomes to wounding in cultured epithelial cells. Mol Biol Cell 11(3):1077–1092

162. Xue M, Thompson P, Kelso I, Jackson C (2004) Activated protein C stimulates proliferation, migration and wound closure, inhibits apoptosis and upregulates MMP-2 activity in cultured human keratinocytes. Exp Cell Res 299(1):119–127

163. Bullard KM, Lund L, Mudgett JS, Mellin TN et al (1999) Impaired wound contraction in stromelysin-1-deficient mice. Ann Surg 230(2):260–265

164. Di Colandrea T, Wang L, Wille J, D'Armiento J et al (1998) Epidermal expression of collagenase delays wound-healing in transgenic mice. J Invest Dermatol 111(6):1029–1033

165. Salonurmi T, Parikka M, Kontusaari S, Pirila E et al (2004) Overexpression of TIMP-1 under the MMP-9 promoter interferes with wound healing in transgenic mice. Cell Tissue Res 315(1):27–37

166. Geer DJ, Andreadis ST (2003) A novel role of fibrin in epidermal healing: plasminogen-mediated migration and selective detachment of differentiated keratinocytes. J Invest Dermatol 121(5):1210–1216

167. Bechtel MJ, Reinartz J, Rox JM, Inndorf S et al (1996) Upregulation of cell-surface-associated plasminogen activation in cultured keratinocytes by interleukin-1 beta and tumor necrosis factor-alpha. Exp Cell Res 223(2):395–404

168. Cale JM, Lawrence DA (2007) Structure-function relationships of plasminogen activator inhibitor-1 and its potential as a therapeutic agent. Curr Drug Targets 8(9):971–981

169. Maquerlot F, Galiacy S, Malo M, Guignabert C et al (2006) Dual role for plasminogen activator inhibitor type 1 as soluble and as matricellular regulator of epithelial alveolar cell wound healing. Am J Pathol 169(5):1624–1632

170. Pivarcsi A, Kemeny L, Dobozy A (2004) Innate immune functions of the keratinocytes. A review. Acta Microbiol Immunol Hung 51(3):303–310

171. Tokumaru S, Sayama K, Shirakata Y, Komatsuzawa H et al (2005) Induction of keratinocyte migration via transactivation of the epidermal growth factor receptor by the antimicrobial peptide LL-37. J Immunol 175(7):4662–4668

172. Loots MA, Lamme EN, Zeegelaar J, Mekkes JR et al (1998) Differences in cellular infiltrate and extracellular matrix of chronic diabetic and venous ulcers versus acute wounds. J Invest Dermatol 111(5):850–857

173. Yager DR, Zhang LY, Liang HX, Diegelmann RF et al (1996) Wound fluids from human pressure ulcers contain elevated matrix metalloproteinase levels and activity compared to surgical wound fluids. J Invest Dermatol 107(5):743–748

174. Ashcroft GS, Horan MA, Ferguson MW (1997) The effects of ageing on wound healing: immunolocalisation of growth factors and their receptors in a murine incisional model. J Anat 190(Pt 3):351–365

175. Pierce GF, Tarpley JE, Tseng J, Bready J et al (1995) Detection of platelet-derived growth factor (PDGF)-AA in actively healing human wounds treated with recombinant PDGF-BB and absence of PDGF in chronic nonhealing wounds. J Clin Invest 96(3):1336–1350

176. Wenk J, Brenneisen P, Meewes C, Wlaschek M et al (2001) UV-induced oxidative stress and photoaging. Curr Probl Dermatol 29:83–94

177. Morasso MI, Tomic-Canic M (2005) Epidermal stem cells: the cradle of epidermal determination, differentiation and wound healing. Biol Cell 97(3):173–183

Chapter 2
Role of Hyaluronan in Wound Healing

Abstract Hyaluronan is a negatively charged high molecular weight connective tissue glycosaminoglycan (GAG). As well as heparin sulfate proteoglycan, hyaluronan is a major polysaccharide found in extra cellular matrix (ECM). Hyaluronan is mainly synthesized by mesenchymal cells and extruded into ECM in coordination with synthesis. Hyaluronan molecules form a continuous but porous meshwork structure. This property of hyaluronan may contribute to the hydrated microenvironment at sites of synthesis. Together with passive functions such as space filling molecule, hyaluronan interacts with cell surface receptors, and regulates cell function by activating intracellular signaling pathway. Hyaluronan is taken up by cell surface receptors for intracellular degradation. The degradation occurs in stepwise fashion by distinct enzymes, hyaluronidase-1 (HYAL1) and HYAL2. Hyaluronan fragment has distinct functions that are not found in normal high molecular weight hyaluronan. High molecular weight hyaluronan acts as a component of intact ECM and tends to maintain signals that promote normal cellular functions whereas hyaluronan fragment tends to induce cellular differentiation, tissue morphogenesis or tissue defense in response to injury. Consequently, hyaluronan exerts beneficial effects on many steps of wound healing process such as inflammation, reepithelialization and resolution.

Keywords Scar • CD44 • Hyaluronan synthase • Hyaluronidase • RHAMM

2.1 Structure

Hyaluronan is composed of D-glucuronic acid and D-*N*-acetylglucosamine, linked via alternating β-1,4 and β-1,3 glycosidic bonds. Hyaluronan polymer is a repeat of disaccharide units in a linear fashion. Hyaluronan differs in several respects from typical GAGs in that it does not contain any sulfate group or uronic acid residues and it does not covalently attach to core protein [1–3]. There are no deviations from linear disaccharide structure. The number of the disaccharide unit reaches 1,000 or more [3, 4].

Hence, the molecular weight of hyaluronan polymer could reach about 10^7 kDa. In cartilarge, the average hyaluronan size is estimated as 2,000 kDa [5]. In normal synovial fluid, it reaches 7,000 kDa [6].

Hyaluronan conformation is affected by physiological conditions. In solid state, single helical conformation was identified. The number of disaccharide per helical turn varies with the type on the counterion [7]. This helical structure is stabilized with hydrogen bonds linking adjacent sugar residues. In aqueous solution at physiological pH, hyaluronan exists in flexible coiled configuration [1, 8, 9]. In dilute solutions, hyaluronan chain interferes with each other. The volume that is occupied by each hyaluronan molecule expands with water trapped inside the structure [10]. A stretched chain of hyaluronan with a molecular weight of 6,000 kDa would have an approximate length of 15 μm, and a diameter of about 0.5 nm [7]. In more concentrated solutions, the hyaluronan molecules become entangled, forming a continuous but porous meshwork structure. Small molecules such as water and nutrients can diffuse through the hyaluronan solutions. However, the diffusion of macromolecules such as protein is inhibited by the porous network structure. An aqueous medium containing an appropriate concentration of hyaluronan has far higher viscosity in comparison to a solution only containing proteins [11]. The viscosity of the synovial fluid is directly related to the hyaluronan content.

2.2   Biosynthesis

Hyaluronan is mainly synthesized in plasma membrane of fibroblasts and other mesenchymal cells [12, 13]. Hyaluronan is also synthesized in epidermal cells [14]. Although, most GAG are synthesized in the Golgi network within the cells and attached to core proteins. Hyaluronan is not associated to a core protein. It is synthesized at inner surface of plasma membrane by hyaluronan synthase (HAS) [15, 16]. HASs are membrane-bound enzymes that use UDP-α-N-acetyl-D-glucosamine and UDP-α-D-glucuronate as substrates to produce the GAG hyaluronan. Hyaluronan chain grows at the reducing end by addition of sugar residues donated by activated UDP-sugar precursors in the presence of Mg^{2+} or Mn^{2+}. The growing polymer is extruded through the membrane to the outside of the cells. It results in unconstrained polymer growth, and production of high molecular weight pericellular hyaluronan coat. It can foam a complex with exogeneous hyaluronan and hyaluronan binding proteoglycans. Such hyaluronan-dependent coat may be required for migration and proliferation of vascular smooth muscle cells. In some cases, cells release hyaluronan from the cell surface by unknown mechanism.

The presence of HAS is known in human, mice, frogs, bacteria and virus. In human, three kinds of HAS have been identified, designated as HAS1, HAS2 and HAS3 [15, 17]. They are encoded by distinct genes located on different chromosome [18]. All the HAS isoforms are highly homologous with their amino acid sequence and have similar structure. The amino acid identity among them is 55–71%. They contain six to eight transmembrane domains and two membrane associated proteins.

Expression of any mammalian HAS can induce hyaluronan secretion, suggesting that each HAS is capable of hyaluronan synthesis and translocation to extracellular space [17, 19]. All three HAS transfectants showed an enhanced motility, and a significant increase in their confluent cell densities [16]. *In vitro* experiments indicate that HAS1 and HAS2 polymerize hyaluronan up to 2×10^6 Da, whereas HAS3 polymerize hyaluronan in the range of 2×10^5 Da [17]. The expression pattern of HAS isoforms is cell type specific [20]. Human dermal fibroblasts predominantly express HAS1 and HAS2 [21], whereas HAS3 is major HAS in human epidermal keratinocytes [22].

HAS2 gene is not an expression of functional redundancy but rather point at significant difference at their biological function. HAS2 plays important role in embryonic development. Hyaluronan content decreases just before birth and then increases again in the day after birth. Has2-deficient mice show severe cardiac and vascular defects, and die during midgestation (E9.5-E10) [23]. In developing heart, Has2 is required for an epithelial-mesenchymal transformation (EMT) and subsequent migration of endothelial cells. However, Has1 and Has3 knockout is not embryonic lethal, suggesting that they participate in lesser roles [23].

Hyaluronan accumulation is observed in response to tissue injury. The radiation-induced lung injury induces Has2 expression in rats [24]. Epidermal trauma induces Has2 and Has3 transcription in adult mice [25]. Ventilation-induced lung injury promotes HAS3 expression [26].

Several growth factors such as epidermal growth factor (EGF), platelet-derived growth factor (PDGF), IGF-1 and transforming growth factor-β (TGF-β) and cytokines (IL-1) can regulate HAS expression [27–29]. HAS2 mRNA expression is increased by IL-1β and TNF-α in lung fibroblasts [30]. EGF promotes HAS2 expression in rat epidermal keratinocytes [31]. KGF induces transcription of HAS2 and HAS3. The TGF-β-induced expression of HAS1 is dependent on p38 MAPK in human synoviocytes [32]. PDGF increases HAS2 mRNA in mesothelial cells [33]. Hyaluronan biosynthesis is enhanced by protein kinase C (PKC) agonist phorbol ester. Many consensus PKC phosphorylation sites are found in the intracellular loop of the HAS protein, suggesting that the HAS activity may be regulated by PKC-dependent mechanism.

In fibroblasts, the rate of hyaluronan synthesis is also dependent on cell density. At low cell density, hyaluronan biosynthesis is high. At high cell density, cell proliferation is slow down and hyaluronan biosynthesis is shut down.

2.3 Hyaluronan Receptor

About 30% of total body hyaluronan is replaced in each day [34]. In circulation, the half-life of hyaluronan is only a few minutes. This rapid turnover of hyaluronan results from hyaluronan uptake by liver endothelial cells. CD44 [35] and the receptor for hyaluronan-mediated motility (RHAMM) [36, 37] are involved in cellular interaction with hyaluronan. CD44 was first identified as a lymphocyte homing receptor mediating the attachment of circulating lymphocytes to endothelial cells [38]. CD44 positive lymphocytes interact with endothelial hyaluronan, migrate the site of inflammation [39]. In addition, CD44 expression is observed in many types of cells including fibroblasts, keratinocytes, smooth muscle cells, osteoclasts, macrophages, neutrophils and lymphocytes, and is considered as primary receptor for hyaluronan [40]. CD44 is synthesized as single trans-membrane protein. CD44 has the distal and proximal extracellular domain. The distal extracellular domain is responsible for hyaluronan binding. Alternative splicing in proximal domain generates various CD44 isoforms. The transmembrane domain is highly conserved. Another site of alternative splicing is intracellular domain. CD44 binds a subset of heparin-binding growth factors, cytokines and ECM components. Hyaluronan-CD44 interactions are involved in development, inflammation, tumor growth and metastasis [41]. CD44 blocking antibody inhibits fibroblast migration into fibrin matrix [42]. Most CD44 isoforms contain cytoplasmic domain that is responsible for interaction with cytoskelton and intracellular signaling proteins. CD44 phosphorylation is essential for a control mechanism for CD44-mediated cell motility. CD44 phosphrylation mutants can attach hyaluronan as wild type CD44. However, they are defective in hyaluronan-mediated fibroblast migration [43]. PKC-dependent phosphorylation of intracellular domain regulates its interaction with cytoskeletal linker protein ezrin *in vivo* [44]. In epithelial cells, hyaluronan regulates TGF-β signaling by interacting CD44 [45, 46]. Smad1 can interact with cytoplasmic domain of CD44 in BMP-7 signaling pathway [47]. As CD44 was first identified as a lymphocyte homing receptor, the role of CD44 on lymphocyte migration is well known. Hyaluronan regulates migration of neutrophils into inflamed tissues by CD44 dependent mechanism [48]. Hyaluronan binding to CD44 induces IL-6 production in neutrophils, and induces IL-1 and TNF-α in monocytes [49]. Ligand binding of CD44 augments phagocytic activity of macrophages [50].

RHAMM (CD168) is another functional hyaluronan receptor. RHAMM was characterized as hyaluronan receptor complex that mediates hyaluronan-induced cell migration [51–53]. RHAMM is a cytoskeleton-associated protein that lacks a signal peptide. RHAMM is released as an extracellular protein most likely by unconventionally export mechanisms, and acts as coreceptor of CD44. RHAMM binds to hyaluronan via a short amino acid sequence containing multiple basic amino acids. Expression of RHAMM is increased in migrating fibroblasts. RHAMM-deficient mice are viable. However, *in vitro* wound healing is abrogated in fibroblasts derived from RHAMM-deficient mice [54]. RHAMM is involved in hyaluronan-enhanced smooth muscle cell migration. An antibody against RHAMM inhibits hyaluronan binding to RHAMM and migration of smooth muscle cells in response to injury [55]. Hyaluronan oligosaccharides are potential stimulators to angiogenesis via RHAMM-dependent mechanism in wound healing [56]. Overexpression of RHAMM in fibroblasts results in Ras-dependent transformation [57]. RHAMM activates a number of protein kinases including ERK1/2, c-Src, focal adhesion kinase (FAK) and Rho-kinase [58, 59]. These observations indicate that RHAMM acts as a functional hyaluronan receptor, and its expression is requited for hyaluronan-enhanced migration of fibroblasts and smooth muscle cells.

However, RHAMM does not contain any protein motif involved in regulation of cell migration and cell cycle. One possible explanation is that RHAMM is a coreceptor for receptors with tyrosine kinase activity. The precise molecular mechanism of RHAMM-mediated cell growth and migration remains to be elucidated.

Toll-like receptors (TLR) not only play a role in the recognition of pathogens and in the initiation of inflammatory response but also have a role in noninfectious disease pathogenesis. Hyaluronan is known as a component of group A and C of *Streptococcus* [4]. High molecular weight hyaluronan protects lung alveolar epithelial cells from bleomycin-induced apoptosis by TLR-dependent mechanism [60]. Hyaluronan is recognized by TLR4, MD-2 and CD44 complex, whereas LPS is recognized by TLR4, MD-2 and CD14 complex. As a result, hyaluronan and LPS induce different sets of gene expression [61]. TLR4 is involved in maturation of dendritic cells that induced by hyaluronan oligosaccharide fragment [62].

LYVE-1 (lymphatic vessel endothelial hyaluronan receptor) is a single transmembrane protein having a single hyaluronan-binding domain in N-terminal [63]. LYVE-1 is involved in hyaluronan transport from tissue to lymph by uptaking hyaluronan via lymphatic endothelial cells. However, hyaluronan homeostasis is not abrogated in LYVE-1-deficient mice, suggesting that loss of LYEL-1 function may be compensated by other hyaluronan receptors [64].

Tissue-specific hyaluronan binding proteins such as brevican and neurocan regulate neuronal development and brain tumor metastasis.

2.4 Mammalian Hyaluronidase

Hyaluronan polymer is cleaved into small fragments by hyaluronidase. Hyaluronan degradation occurs stepwise by different hyaluronidases. In human, six hyaluronidases are identified. Only three proteins (HYAL1, HYAL2 and HYAL3) are widely expressed in somatic tissue. HYAL1 is first identified hyaluronidase present in plasma and urine [65, 66]. It is also a lysosomal enzyme, and degrades hyaluronan to small disaccharide, mainly tetrasaccharide. Although HYAL1-deficient mice are viable and fertile, deletion of HYAL1 is observed in lung, head, neck tumor, suggesting that HYAL1 exerts a tumor suppressor effect.

HYAL2 is a GPI-anchored protein [67]. A portion of HYAL2 is released into extracellular space by unknown mechanism [68]. HYAL2 also exists as lysosomal protein. HYAL2 cleaves only high molecular weight hyaluronan into intermediate sized hyaluronan fragment (<20 kDa) [69]. HYAL2-deficient mice show elevation of plasma hyaluronidase activity and hyaluronan concentration [70]. Furthermore, HYAL2 acts as a coreceptor of hyaluronan at cell surface along with CD44 [71]. HYAL3 expression is observed in testis, bone marrow, testis and kidney [66]. Its structure is similar to HYAL1 and HYAL2. Together with HYAL2, HYAL3 expression is increased by inflammatory cytokines [72]. However, no enzymatic activity is detected using the available hyaluronidase assays [73]. Accumulation of hyaluronan is not observed in HYAL3-deficient mice [74].

HYAL4 is detected in placenta and skeletal muscle [66]. The enzymatic activity of HYAL4 is weak [75]. PH-20 (HYAL5) is highly expressed in testis and breast tumors. PH-20 hyaluronidase activity is required for penetration of sperm through the cumulus mass that surrounds the ovum [76]. PHYAL1 is known as a pseudogene that does not encode any protein.

2.5   Hyaluronan Fragment

Hyaluronan plays different roles depending on its size [4, 77–80]. In normal state, hyaluronan exists as high molecular weight polymer (10^6–10^7 Da). High molecular weight hyaluronan tends to maintain signals that promote normal cellular functions. High molecular weight hyaluronan has anti-inflammatory and immuno-suppressive properties [81]. HYAL2 digestion of hyaluronan results in lower molecular weight hyaluronan of 20 kDa. Fragmentation of hyaluronan occurs in some pathogenetic conditions such as tissue injury and inflammation. Hyaluronan fragments tend to induce cellular differentiation, tissue morphogenesis or tissue defense in response to injury. The hyaluronan fragments are internalized by receptor-mediated endocytosis. They are potent stimulator of inflammatory cytokines [82]. The stress induced hyaluronan fragmentation activates corresponding immunological pathways. In normal lung, most hyaluronan is a high molecular weight form. In response to acute lung injury, hyaluronan chain smaller than 500 kDa can be detected at concentration sufficient to induce inflammation [83]. Hyaluronan fragments induce maturation of dendritic cells [84]. Langerhans cells are epidermis-specific debtritic cells and critical for initiation of cutaneous immune response. Administration of hyaluronan blocking peptide antagonizes hapten-induced maturation of Langerhans cells *in vivo* [85]. Flavonoids, including silybin, apigenin, kaempferol, luteloin are known as competitive inhibitor of hyaluronidases [86]. Some of them show anti-inflammatory activity.

Hyaluronan fragment promotes migration of endothelial cells and angiogenesis, whereas high molecular weight hyaluronan inhibits migration of endothelial cells and angiogenesis [87]. The effect of hyaluronan fragment is dependent on both CD44 and RHAMM [88–90]. Smaller hyaluronan fragments (6–20 kDa) deliver maturation signal to dendritic cells [91, 92]. Smaller hyaluronan fragment inhibits airway response that induced by neutrophil elastase in sheep. Very small hyaluronan oligosaccharides have unique biological activities. Hyaluronan oligosaccharide (3–10 saccharide) inhibits anchorage-independent tumor growth [93]. Tetrasaccharides, predominant product of HYAL1, promotes the expression of heat shock proteins [94]. It has anti-apoptotic effect. It also indices angiogenesis in chick chorioallantoic membrane to loosen or destabilize cell-cell interaction [95]. Smaller molecular weight hyaluronan enhances IFN-γ synthesis by NF-κB dependent mechanism.

2.6 Role on Cancer

The ECM components serve as a pliable scaffold that enables cell proliferation and migration. In addition, ECM amplifies growth factor signals that synergistically promote cell proliferation and migration. The transient accumulation of hyaluronan is observed in adult tissues undergo rapid remodeling, such as in inflammation, and cancer. Most types of human cancers require supportive element that promotes the growth of malignant cells. Usually, the supportive elements consist of connective tissue containing vascular structures [96]. Hyaluronan has an ability to control cell growth and migration in tumor development and metastasis in various cancer tissues [97, 98]. Hyaluronan levels are increased in most malignant solid tumor [99]. The level of hyaluronan in peritumoral stroma and malignant cells correlates with the extent of breast cancer spreading and metastasis [100]. Melanoma cells selected for high expression of hyaluronan are more metastatic when injected into nude mice than cells that expressed low amounts of hyaluronan [101]. A numbers of cancers are associated with elevation of HAS expression. Overproduction of hyaluronan promotes anchorage-independent growth of fibrosarcoma and mammary carcinoma [102]. According to studies using transgenic mouse, HAS2 overexpression augments tumor development, angiogenesis and lymphangiogenesis [103, 104]. Increased hyaluronan production promotes growth and invasiveness of normal canine kidney and MCF-10A human mammary epithelial cells [105]. Overexpression of HAS3 promotes the growth of prostate cancer cells [106].

Furthermore, hyaluronan facilitates tumor growth by opening up spaces for cell migration. Hyaluronan affects tumor phenotype through binding to hyaluronan receptors. Both CD44 and RHAMM is involved in hyaluronan-induced tumor cell migration. Hyaluronan-induced clustering of CD44 and overexpression of RHAMM promotes the tumor inversiveness [57, 107].

2.7 Hyaluronan in Dermis and Epidermis

The largest amount of hyaluronan (about 50% of the total body) exists in skin tissue. Hyaluronan concentration is about 0.5 mg/g wet tissue in dermis, and 0.1 mg/g wet tissue in epidermis. Hyaluronan has a rapid turnover with a half-life of about 1 day in dermis and epidermis [108]. Hyaluronan is major space filling molecule in dermis. Fibroblasts express all three isoforms of HAS (HAS1, HAS2 and HAS3). Hyaluronan interacts with collagen, fibronectin and laminins [109, 110]. In addition, hyaluronan is regarded as integral part of elastin fiber and collagen microfibrils [111].

The epidermis is a stratified epithelium, composed of proliferating basal and differentiated subbasal keratinocytes. After cell division, daughter keratinocytes are retained in basal layer or migrate upward to undergo differentiation. Hyaluronan is major GAG in epidermis and fills narrow extracellular space between basal and subbasal keratinocytes [112]. As epidermis is a closed compartment, and penetration of macromolecules from dermis is limited by stratum corneum and basal lamina. Synthesis and catabolism of epidermal hyaluronan is likely to be mediated by keratinocytes [113]. All three HAS is also expressed in keratinocytes. Most of newly synthesized hyaluronan is high molecular weight ($>2 \times 10^6$) in keratinocyte monolayer culture and normal epidermis. The epithelial localization of CD44, a primary hyaluronan receptor, overlaps with that of hyaluronan [114]. The endocytosed hyaluronan has a half-life of 2–3 h. The expression of hyaluronan is not detected in the terminal differentiated keratinocytes.

2.8   Role on Wound Healing

Hyaluronan exerts beneficial effects on wound healing. Topically applied hyaluronan accelerates cutaneous wound healing in rats [115] and hamsters [116]. Fibroblasts release hyaluronan in response to tissue injury [26, 33, 117]. Hyaluronan acts as a promoter of early inflammation, which is crucial in the whole skin wound-healing process. Hyaluronan promotes production of pro-inflammatory cytokines such as IL-1β, IL-8 and TNF-α [118–120]. Hyaluronan accumulation is critical for formation of granulation tissue during wound healing process. Hyaluronan makes space that allows the infiltration of neutrophils and macrophages in the plasma clot. Hyaluronan interacts with collagen, fibronectin and fibrinogen, and promotes fibrin polymerization and clot formation [121, 122]. The correlation of hyaluronan synthesis and fibroblast migration is compatible with the promoting effect of hyaluronan on fibroblast migration during embryonic tissue development and in regenerative processes [123, 124].

In granulation tissue, differentiation of fibroblasts to myofibroblasts is associated with hyaluronan accumulation through reduced hyaluronan catabolism [125]. Myofibroblasts play important roles in wound contraction and fibrosis development. They have intermediate phenotype between fibroblasts and smooth muscle cells, characterized by the expression of α-smooth muscle actin (SMA) [126, 127]. Accumulation of hyaluronan pericellular coats is special feature of myofibroblast differentiation [128]. The assembly of pericellular hyaluronan coat is correlated with inflammation, wound healing, tumor invasion and EMT by increasing cell proliferation and migration [105, 129–132]. HAS inhibitor antagonizes TGF-β-induced hyaluronan accumulation and α-SMA expression [133]. Taken together, hyaluronan is critical for TGF-β-induced myofibroblast differentiation and augments the expression of the differentiated phenotype.

The effect of hyaluronan on fibroblast migration is mediated by hyaluronan receptors and hyaluronan binding proteins. During wound healing process, CD44 is a principal receptor that mediates fibroblast interaction with hyaluronan. CD44 expression can be detected uniformly cell surface, especially in filopodia and lamellipodia. CD44 is required for invasion of fibroblasts into fibronectin/fibrin gels [134]. Human lung fibroblasts isolated from patients with acute alveolar fibrosis use

CD44 to invade into a fibrin matrix [42]. In CD44-deficient fibroblasts, the number of actin stress fiber and focal adhesion is decreased [135]. TGF-β-induced proliferation and collagen synthesis is reduced in the CD44-deficient fibroblasts [136]. CD44 mediates adhesion of the lung fibroblasts to fibrin clot components such as hyaluronan, fibronectin and fibrinogen. RHAMM is not highly expressed in normal tissue homeostasis. The expression of RHAMM is increased in response to injury. RHAMM-deficient fibroblasts fail to repair wounds [54]. They are defective in CD44-mediated activation of ERK1/2 [137], suggesting that RHAMM acts as coreceptor for CD44 and mediates CD44-induced ERK1/2 activation.

Hyaluronan accumulation can be observed in area spreads from the wounded margins, suggesting that some growth factors released from the wounded site increase hyaluronan synthesis [138]. In injured epidermis, heparin-binding EGF-like growth factor (HB-EGF) released from keratinocytes promotes hyaluronan synthesis in a paracrine manner.

Hyaluronan protects fibroblast damages induced by reactive oxygen species (ROS). Hyaluronan acts as ROS scavenger and protects articular tissues [139]. ROS is released from macrophages and polymorphonuclear leukocytes (PMNs) in response to injury. Excess ROS induces prolonged inflammation and degrades ECM components such as collagen, laminin and hyaluronan *in vitro* [140]. ROS-induced hyaluronan degradation is achieved by direct action of ROS [141], or activation of hyaluronidase [142]. Hyalironan degradation exaggerates inflammation at the site of injury as hyaluronan fragment augments inflammatory response. ROS scavengers (e.g. superoxide dismutase, catalase and desferrioxamine) are known to reduce hyaluronan degradation [143], suggesting that inhibition of oxidative hyaluronan fragmentation probably represents one mechanism by which SOD scavenger inhibits inflammation in response to injury.

Scar is a fibrous tissue that develops in normal wound healing process after injury [144–146]. However, extensive scar formation, which characterized by persistent and excessive fibroblast proliferation and collagen deposition, inhibits wound resolution. Hyaluronan is likely to contribute scarless wound healing. In adult wounds, scar formation is associated with disappearance of high molecular weight hyaluronan. In initial stage of wound healing, hyaluronan is accumulated in fibrin clots. This fibronectin/fibrin rich provisional matrix is replaced by collagen-rich matrix, along with the degradation of hyaluronan by hyaluronidases [146, 147]. It returns to baseline level in 21 days after wounding. The lack of scaring and inflammation is a special feature of fetal skin wound healing [147]. In fetal wound, hyaluronan accumulation persists throughout the wound healing process [148]. Perhaps, scarring plays a much less prominent role in fetal wound healing. Unlike adult fibroblasts, hyaluronan production does not attenuated in confluent fetal fibroblasts. They continue to produce the same amount of hyaluronan in confluent state. The prolonged hyaluronan accumulation in fetal wounds likely to provide an environment that is suitable for fibroblast migration [147, 149]. The regulation of collagen synthesis is obviously most important mechanism for scarless repair. Hyaluronan promotes type-III collagen expression in human dermal fibroblasts [150]. Fetal skin has a higher ratio of type-III to type-I collagen than adult skin [151, 152]. Type III-deficient mice show significantly more scar tissue area compared to wild-type mice [153]. Hyaluronan is likely to provide the fetal like environment which known to promote scarless wound healing. The underlying mechanisms by which hyaluronan inhibits scar formation during the healing process remain unknown.

Hyaluronan also has crucial functions in the reepithelialization due to several properties. In epidermis, hyaluronan concentration increases up to sevenfold in response to injury [25]. Epidermal keratinocytes adjacent to wounded area migrate into wound area to cover the damaged area and reconstruct a new epithelium. The process is called reepithelialization. Factors that enhance hyaluronan synthesis (EGF, KGF and retinoic acid) tend to promote keratinocyte proliferation, whereas factors that inhibit hyaluronan synthesis (TGF-β and hydrocortisone) inhibit keratinocyte proliferation [14, 138, 154, 155]. A hyaluronan synthesis inhibitor (4-Methylumbelliferone) prevents EGF-induced keratinocyte proliferation [156]. Removal of hyaluronan from rat epithelial keratinocytes accelerates keratinocyte differentiation [157]. Suppression of CD44 expression in skin keratinocytes results in abnormal hyaluronan accumulation in the superficial dermis [37]. Skin elasticity, local inflammatory response, and tissue repair were also impaired by the suppression of CD44 expression. These defects are accompanied by abrogated keratinocyte proliferation in response to mitogen and growth factors. These lines of observations suggest that hyaluronan supports proliferative phenotype of dermal keratinocytes, and inhibits the expression of differentiated phenotype [138]. The exact role of hyaluronan on keratinocyte proliferation and differentiation is not known. The intercellular accumulation of hyaluronan inhibits adhesion between keratinocytes and making space for cell migration.

The proposed effects of retinoic acid (vitamin A) against skin damage and aging may be correlated with an increase of skin hyaluronan content. Retinoic acid inhibits the terminal differentiation of keratinocytes by hyaluronan-dependent mechanism. It promotes the hyaluronan production in human skin organ culture [158]. The treatment with retinoic acid induces accumulation of hyaluronan in intracellular space, and reduces the number of desmosomes between keratinocytes. Hydrocortisone (cortisol) is popular as an anti-inflammatory agent. Hydrocortisone prevents hyaluronan degradation at all dose [159]. Hydrocortisone, at a level slightly exceeding the physiological concentration (10^{-6} M), decreased hyaluronan synthesis by 50% [160]. Therefore, hydrocortisone stabilizes hyaluronan at all doses and reduces the content of hyaluronan in large doses. Hydrocortisone is widely used as an anti-proliferative agent in skin disease.

Excess accumulation of hyaluronan disrupts epithelial barrier structure, and increases the thickness of epidermis by inhibiting the terminal differentiation of keratinocytes. The rate of hyaluronan synthesis and degradation of hyaluronan should be coupled. However, the mechanism that controls the total amount of epidermal hyaluronan is unknown.

References

1. Laurent TC, Fraser JR (1992) Hyaluronan. FASEB J 6(7):2397–2404
2. Toole BP (2000) Hyaluronan is not just a goo! J Clin Invest 106(3):335–336
3. Stern R (2003) Devising a pathway for hyaluronan catabolism: are we there yet? Glycobiology 13(12):105R–115R
4. Jiang D, Liang J, Noble PW (2011) Hyaluronan as an immune regulator in human diseases. Physiol Rev 91(1):221–264
5. Holmes MW, Bayliss MT, Muir H (1988) Hyaluronic acid in human articular cartilage. Age-related changes in content and size. Biochem J 250(2):435–441
6. Fraser JR, Laurent TC, Laurent UB (1997) Hyaluronan: its nature, distribution, functions and turnover. J Intern Med 242(1):27–33
7. Cowman MK, Matsuoka S (2005) Experimental approaches to hyaluronan structure. Carbohydr Res 340(5):791–809
8. Cleland RL (1970) Ionic polysaccharides. IV. Free-rotation dimensions for disaccharide polymers. Comparison with experiment for hyaluronic acid. Biopolymers 9(7):811–824
9. Scott D, Coleman PJ, Mason RM, Levick JR (1998) Direct evidence for the partial reflection of hyaluronan molecules by the lining of rabbit knee joints during trans-synovial flow. J Physiol 508(Pt 2):619–623
10. Toole BP (2004) Hyaluronan: from extracellular glue to pericellular cue. Nat Rev Cancer 4(7):528–539
11. Stojkovic M, Kolle S, Peinl S, Stojkovic P et al (2002) Effects of high concentrations of hyaluronan in culture medium on development and survival rates of fresh and frozen-thawed bovine embryos produced in vitro. Reproduction 124(1):141–153
12. Prehm P (1990) Release of hyaluronate from eukaryotic cells. Biochem J 267(1):185–189
13. Watanabe K, Yamaguchi Y (1996) Molecular identification of a putative human hyaluronan synthase. J Biol Chem 271(38):22945–22948
14. Pasonen-Seppanen S, Karvinen S, Torronen K, Hyttinen JM et al (2003) EGF upregulates, whereas TGF-beta downregulates, the hyaluronan synthases Has2 and Has3 in organotypic keratinocyte cultures: correlations with epidermal proliferation and differentiation. J Invest Dermatol 120(6):1038–1044
15. Weigel PH, Hascall VC, Tammi M (1997) Hyaluronan synthases. J Biol Chem 272(22):13997–14000
16. Itano N, Atsumi F, Sawai T, Yamada Y et al (2002) Abnormal accumulation of hyaluronan matrix diminishes contact inhibition of cell growth and promotes cell migration. Proc Natl Acad Sci USA 99(6):3609–3614
17. Itano N, Sawai T, Yoshida M, Lenas P et al (1999) Three isoforms of mammalian hyaluronan synthases have distinct enzymatic properties. J Biol Chem 274(35):25085–25092
18. Spicer AP, Seldin MF, Olsen AS, Brown N et al (1997) Chromosomal localization of the human and mouse hyaluronan synthase genes. Genomics 41(3):493–497
19. Spicer AP, McDonald JA (1998) Characterization and molecular evolution of a vertebrate hyaluronan synthase gene family. J Biol Chem 273(4):1923–1932
20. Tien JY, Spicer AP (2005) Three vertebrate hyaluronan synthases are expressed during mouse development in distinct spatial and temporal patterns. Dev Dyn 233(1):130–141
21. Sugiyama Y, Shimada A, Sayo T, Sakai S et al (1998) Putative hyaluronan synthase mRNA are expressed in mouse skin and TGF-beta upregulates their expression in cultured human skin cells. J Invest Dermatol 110(2):116–121
22. Sayo T, Sugiyama Y, Takahashi Y, Ozawa N et al (2002) Hyaluronan synthase 3 regulates hyaluronan synthesis in cultured human keratinocytes. J Invest Dermatol 118(1):43–48
23. Camenisch TD, Spicer AP, Brehm-Gibson T, Biesterfeldt J et al (2000) Disruption of hyaluronan synthase-2 abrogates normal cardiac morphogenesis and hyaluronan-mediated transformation of epithelium to mesenchyme. J Clin Invest 106(3):349–360

24. Li Y, Rahmanian M, Widstrom C, Lepperdinger G et al (2000) Irradiation-induced expression of hyaluronan (HA) synthase 2 and hyaluronidase 2 genes in rat lung tissue accompanies active turnover of HA and induction of types I and III collagen gene expression. Am J Respir Cell Mol Biol 23(3):411–418
25. Tammi R, Pasonen-Seppanen S, Kolehmainen E, Tammi M (2005) Hyaluronan synthase induction and hyaluronan accumulation in mouse epidermis following skin injury. J Invest Dermatol 124(5):898–905
26. Bai KJ, Spicer AP, Mascarenhas MM, Yu L et al (2005) The role of hyaluronan synthase 3 in ventilator-induced lung injury. Am J Respir Crit Care Med 172(1):92–98
27. Suzuki M, Asplund T, Yamashita H, Heldin CH et al (1995) Stimulation of hyaluronan biosynthesis by platelet-derived growth factor-BB and transforming growth factor-beta 1 involves activation of protein kinase C. Biochem J 307(Pt 3):817–821
28. Heldin P, Laurent TC, Heldin CH (1989) Effect of growth factors on hyaluronan synthesis in cultured human fibroblasts. Biochem J 258(3):919–922
29. Li L, Asteriou T, Bernert B, Heldin CH et al (2007) Growth factor regulation of hyaluronan synthesis and degradation in human dermal fibroblasts: importance of hyaluronan for the mitogenic response of PDGF-BB. Biochem J 404(2):327–336
30. Wilkinson TS, Potter-Perigo S, Tsoi C, Altman LC et al (2004) Pro- and anti-inflammatory factors cooperate to control hyaluronan synthesis in lung fibroblasts. Am J Respir Cell Mol Biol 31(1):92–99
31. Pienimaki JP, Rilla K, Fulop C, Sironen RK et al (2001) Epidermal growth factor activates hyaluronan synthase 2 in epidermal keratinocytes and increases pericellular and intracellular hyaluronan. J Biol Chem 276(23):20428–20435
32. Stuhlmeier KM, Pollaschek C (2004) Differential effect of transforming growth factor beta (TGF-beta) on the genes encoding hyaluronan synthases and utilization of the p38 MAPK pathway in TGF-beta-induced hyaluronan synthase 1 activation. J Biol Chem 279(10):8753–8760
33. Jacobson A, Brinck J, Briskin MJ, Spicer AP et al (2000) Expression of human hyaluronan synthases in response to external stimuli. Biochem J 348(Pt 1):29–35
34. Laurent UB, Dahl LB, Reed RK (1991) Catabolism of hyaluronan in rabbit skin takes place locally, in lymph nodes and liver. Exp Physiol 76(5):695–703
35. Culty M, Nguyen HA, Underhill CB (1992) The hyaluronan receptor (CD44) participates in the uptake and degradation of hyaluronan. J Cell Biol 116(4):1055–1062
36. Hua Q, Knudson CB, Knudson W (1993) Internalization of hyaluronan by chondrocytes occurs via receptor-mediated endocytosis. J Cell Sci 106(Pt 1):365–375
37. Kaya G, Rodriguez I, Jorcano JL, Vassalli P et al (1997) Selective suppression of CD44 in keratinocytes of mice bearing an antisense CD44 transgene driven by a tissue-specific promoter disrupts hyaluronate metabolism in the skin and impairs keratinocyte proliferation. Genes Dev 11(8):996–1007
38. Ponta H, Sherman L, Herrlich PA (2003) CD44: from adhesion molecules to signalling regulators. Nat Rev Mol Cell Biol 4(1):33–45
39. Nandi A, Estess P, Siegelman MH (2000) Hyaluronan anchoring and regulation on the surface of vascular endothelial cells is mediated through the functionally active form of CD44. J Biol Chem 275(20):14939–14948
40. Sherman L, Sleeman J, Herrlich P, Ponta H (1994) Hyaluronate receptors: key players in growth, differentiation, migration and tumor progression. Curr Opin Cell Biol 6(5):726–733
41. Lesley J, Hyman R, Kincade PW (1993) CD44 and its interaction with extracellular matrix. Adv Immunol 54:271–335
42. Svee K, White J, Vaillant P, Jessurun J et al (1996) Acute lung injury fibroblast migration and invasion of a fibrin matrix is mediated by CD44. J Clin Invest 98(8):1713–1727
43. Peck D, Isacke CM (1996) CD44 phosphorylation regulates melanoma cell and fibroblast migration on, but not attachment to, a hyaluronan substratum. Curr Biol 6(7):884–890

44. Legg JW, Lewis CA, Parsons M, Ng T et al (2002) A novel PKC-regulated mechanism controls CD44 ezrin association and directional cell motility. Nat Cell Biol 4(6):399–407

45. Ito T, Williams JD, Fraser D, Phillips AO (2004) Hyaluronan attenuates transforming growth factor-beta1-mediated signaling in renal proximal tubular epithelial cells. Am J Pathol 164(6):1979–1988

46. Ito T, Williams JD, Fraser DJ, Phillips AO (2004) Hyaluronan regulates transforming growth factor-beta1 receptor compartmentalization. J Biol Chem 279(24):25326–25332

47. Peterson LF, Wang Y, Lo MC, Yan M et al (2007) The multi-functional cellular adhesion molecule CD44 is regulated by the 8;21 chromosomal translocation. Leukemia 21(9): 2010–2019

48. Alstergren P, Zhu B, Glogauer M, Mak TW et al (2004) Polarization and directed migration of murine neutrophils is dependent on cell surface expression of CD44. Cell Immunol 231(1–2):146–157

49. Sconocchia G, Campagnano L, Adorno D, Iacona A et al (2001) CD44 ligation on peripheral blood polymorphonuclear cells induces interleukin-6 production. Blood 97(11):3621–3627

50. Vivers S, Dransfield I, Hart SP (2002) Role of macrophage CD44 in the disposal of inflammatory cell corpses. Clin Sci (Lond) 103(5):441–449

51. Yang B, Zhang L, Turley EA (1993) Identification of two hyaluronan-binding domains in the hyaluronan receptor RHAMM. J Biol Chem 268(12):8617–8623

52. Hardwick C, Hoare K, Owens R, Hohn HP et al (1992) Molecular cloning of a novel hyaluronan receptor that mediates tumor cell motility. J Cell Biol 117(6):1343–1350

53. Turley EA, Austen L, Vandeligt K, Clary C (1991) Hyaluronan and a cell-associated hyaluronan binding protein regulate the locomotion of ras-transformed cells. J Cell Biol 112(5):1041–1047

54. Tolg C, Poon R, Fodde R, Turley EA et al (2003) Genetic deletion of receptor for hyaluronan-mediated motility (Rhamm) attenuates the formation of aggressive fibromatosis (desmoid tumor). Oncogene 22(44):6873–6882

55. Savani RC, Wang C, Yang B, Zhang S et al (1995) Migration of bovine aortic smooth muscle cells after wounding injury. The role of hyaluronan and RHAMM. J Clin Invest 95(3):1158–1168

56. Gao F, Yang CX, Mo W, Liu YW et al (2008) Hyaluronan oligosaccharides are potential stimulators to angiogenesis via RHAMM mediated signal pathway in wound healing. Clin Invest Med 31(3):E106–116

57. Hall CL, Yang B, Yang X, Zhang S et al (1995) Overexpression of the hyaluronan receptor RHAMM is transforming and is also required for H-ras transformation. Cell 82(1):19–26

58. Hall CL, Wang C, Lange LA, Turley EA (1994) Hyaluronan and the hyaluronan receptor RHAMM promote focal adhesion turnover and transient tyrosine kinase activity. J Cell Biol 126(2):575–588

59. Hamilton SR, Fard SF, Paiwand FF, Tolg C et al (2007) The hyaluronan receptors CD44 and Rhamm (CD168) form complexes with ERK1,2 that sustain high basal motility in breast cancer cells. J Biol Chem 282(22):16667–16680

60. Jiang D, Liang J, Fan J, Yu S et al (2005) Regulation of lung injury and repair by Toll-like receptors and hyaluronan. Nat Med 11(11):1173–1179

61. Taylor KR, Yamasaki K, Radek KA, Di Nardo A et al (2007) Recognition of hyaluronan released in sterile injury involves a unique receptor complex dependent on Toll-like receptor 4, CD44, and MD-2. J Biol Chem 282(25):18265–18275

62. Termeer C, Benedix F, Sleeman J, Fieber C et al (2002) Oligosaccharides of Hyaluronan activate dendritic cells via toll-like receptor 4. J Exp Med 195(1):99–111

63. Jackson DG (2009) Immunological functions of hyaluronan and its receptors in the lymphatics. Immunol Rev 230(1):216–231

64. Gale NW, Prevo R, Espinosa J, Ferguson DJ et al (2007) Normal lymphatic development and function in mice deficient for the lymphatic hyaluronan receptor LYVE-1. Mol Cell Biol 27(2):595–604

65. Csoka AB, Frost GI, Wong T, Stern R (1997) Purification and microsequencing of hyaluronidase isozymes from human urine. FEBS Lett 417(3):307–310
66. Csoka AB, Scherer SW, Stern R (1999) Expression analysis of six paralogous human hyaluronidase genes clustered on chromosomes 3p21 and 7q31. Genomics 60(3):356–361
67. Rai SK, Duh FM, Vigdorovich V, Danilkovitch-Miagkova A et al (2001) Candidate tumor suppressor HYAL2 is a glycosylphosphatidylinositol (GPI)-anchored cell-surface receptor for jaagsiekte sheep retrovirus, the envelope protein of which mediates oncogenic transformation. Proc Natl Acad Sci USA 98(8):4443–4448
68. Mullegger J, Lepperdinger G (2002) Degradation of hyaluronan by a Hyal2 type hyaluronidase affects pattern formation of vitelline vessels during embryogenesis of Xenopus laevis. Mech Dev 111(1–2):25–35
69. Lepperdinger G, Strobl B, Kreil G (1998) HYAL2, a human gene expressed in many cells, encodes a lysosomal hyaluronidase with a novel type of specificity. J Biol Chem 273(35):22466–22470
70. Jadin L, Wu X, Ding H, Frost GI et al (2008) Skeletal and hematological anomalies in HYAL2-deficient mice: a second type of mucopolysaccharidosis IX? FASEB J 22(12): 4316–4326
71. Knudson W, Chow G, Knudson CB (2002) CD44-mediated uptake and degradation of hyaluronan. Matrix Biol 21(1):15–23
72. Monzon ME, Manzanares D, Schmid N, Casalino-Matsuda SM et al (2008) Hyaluronidase expression and activity is regulated by pro-inflammatory cytokines in human airway epithelial cells. Am J Respir Cell Mol Biol 39(3):289–295
73. Csoka AB, Frost GI, Stern R (2001) The six hyaluronidase-like genes in the human and mouse genomes. Matrix Biol 20(8):499–508
74. Atmuri V, Martin DC, Hemming R, Gutsol A et al (2008) Hyaluronidase 3 (HYAL3) knockout mice do not display evidence of hyaluronan accumulation. Matrix Biol 27(8):653–660
75. Kaneiwa T, Mizumoto S, Sugahara K, Yamada S (2010) Identification of human hyaluronidase-4 as a novel chondroitin sulfate hydrolase that preferentially cleaves the galactosaminidic linkage in the trisulfated tetrasaccharide sequence. Glycobiology 20(3):300–309
76. Gmachl M, Sagan S, Ketter S, Kreil G (1993) The human sperm protein PH-20 has hyaluronidase activity. FEBS Lett 336(3):545–548
77. Horton MR, McKee CM, Bao C, Liao F et al (1998) Hyaluronan fragments synergize with interferon-gamma to induce the C-X-C chemokines mig and interferon-inducible protein-10 in mouse macrophages. J Biol Chem 273(52):35088–35094
78. McKee CM, Penno MB, Cowman M, Burdick MD (1996) Hyaluronan (HA) fragments induce chemokine gene expression in alveolar macrophages. The role of HA size and CD44. J Clin Invest 98(10):2403–2413
79. Noble PW, McKee CM, Cowman M, Shin HS (1996) Hyaluronan fragments activate an NF-kappa B/I-kappa B alpha autoregulatory loop in murine macrophages. J Exp Med 183(5):2373–2378
80. McKee CM, Lowenstein CJ, Horton MR, Wu J et al (1997) Hyaluronan fragments induce nitric-oxide synthase in murine macrophages through a nuclear factor kappaB-dependent mechanism. J Biol Chem 272(12):8013–8018
81. Stuhlmeier KM (2006) Aspects of the biology of hyaluronan, a largely neglected but extremely versatile molecule. Wien Med Wochenschr 156(21–22):563–568
82. Noble PW (2002) Hyaluronan and its catabolic products in tissue injury and repair. Matrix Biol 21(1):25–29
83. Mascarenhas MM, Day RM, Ochoa CD, Choi WI et al (2004) Low molecular weight hyaluronan from stretched lung enhances interleukin-8 expression. Am J Respir Cell Mol Biol 30(1):51–60
84. Tesar BM, Jiang D, Liang J, Palmer SM et al (2006) The role of hyaluronan degradation products as innate alloimmune agonists. Am J Transplant 6(11):2622–2635

85. Mummert ME, Mohamadzadeh M, Mummert DI, Mizumoto N et al (2000) Development of a peptide inhibitor of hyaluronan-mediated leukocyte trafficking. J Exp Med 192(6): 769–779

86. Kuppusamy UR, Khoo HE, Das NP (1990) Structure-activity studies of flavonoids as inhibitors of hyaluronidase. Biochem Pharmacol 40(2):397–401

87. Montesano R, Kumar S, Orci L, Pepper MS (1996) Synergistic effect of hyaluronan oligosaccharides and vascular endothelial growth factor on angiogenesis in vitro. Lab Invest 75(2):249–262

88. Gao F, Cao M, Yang C, He Y et al (2006) Preparation and characterization of hyaluronan oligosaccharides for angiogenesis study. J Biomed Mater Res B Appl Biomater 78(2):385–392

89. Savani RC, Cao G, Pooler PM, Zaman A et al (2001) Differential involvement of the hyaluronan (HA) receptors CD44 and receptor for HA-mediated motility in endothelial cell function and angiogenesis. J Biol Chem 276(39):36770–36778

90. Slevin M, Kumar S, Gaffney J (2002) Angiogenic oligosaccharides of hyaluronan induce multiple signaling pathways affecting vascular endothelial cell mitogenic and wound healing responses. J Biol Chem 277(43):41046–41059

91. Termeer CC, Hennies J, Voith U, Ahrens T et al (2000) Oligosaccharides of hyaluronan are potent activators of dendritic cells. J Immunol 165(4):1863–1870

92. Termeer C, Sleeman JP, Simon JC (2003) Hyaluronan–magic glue for the regulation of the immune response? Trends Immunol 24(3):112–114

93. Ghatak S, Misra S, Toole BP (2002) Hyaluronan oligosaccharides inhibit anchorage-independent growth of tumor cells by suppressing the phosphoinositide 3-kinase/Akt cell survival pathway. J Biol Chem 277(41):38013–38020

94. Xu H, Ito T, Tawada A, Maeda H et al (2002) Effect of hyaluronan oligosaccharides on the expression of heat shock protein 72. J Biol Chem 277(19):17308–17314

95. West DC, Hampson IN, Arnold F, Kumar S (1985) Angiogenesis induced by degradation products of hyaluronic acid. Science 228(4705):1324–1326

96. Sironen RK, Tammi M, Tammi R, Auvinen PK et al (2011) Hyaluronan in human malignancies. Exp Cell Res 317(4):383–391

97. Toole BP, Biswas C, Gross J (1979) Hyaluronate and invasiveness of the rabbit V2 carcinoma. Proc Natl Acad Sci USA 76(12):6299–6303

98. Turley EA, Bowman P, Kytryk MA (1985) Effects of hyaluronate and hyaluronate binding proteins on cell motile and contact behaviour. J Cell Sci 78:133–145

99. Toole BP, Wight TN, Tammi MI (2002) Hyaluronan-cell interactions in cancer and vascular disease. J Biol Chem 277(7):4593–4596

100. Auvinen P, Tammi R, Parkkinen J, Tammi M et al (2000) Hyaluronan in peritumoral stroma and malignant cells associates with breast cancer spreading and predicts survival. Am J Pathol 156(2):529–536

101. Zhang L, Underhill CB, Chen L (1995) Hyaluronan on the surface of tumor cells is correlated with metastatic behavior. Cancer Res 55(2):428–433

102. Kosaki R, Watanabe K, Yamaguchi Y (1999) Overproduction of hyaluronan by expression of the hyaluronan synthase Has2 enhances anchorage-independent growth and tumorigenicity. Cancer Res 59(5):1141–1145

103. Koyama H, Kobayashi N, Harada M, Takeoka M et al (2008) Significance of tumor-associated stroma in promotion of intratumoral lymphangiogenesis: pivotal role of a hyaluronan-rich tumor microenvironment. Am J Pathol 172(1):179–193

104. Koyama H, Hibi T, Isogai Z, Yoneda M et al (2007) Hyperproduction of hyaluronan in neu-induced mammary tumor accelerates angiogenesis through stromal cell recruitment: possible involvement of versican/PG-M. Am J Pathol 170(3):1086–1099

105. Zoltan-Jones A, Huang L, Ghatak S, Toole BP (2003) Elevated hyaluronan production induces mesenchymal and transformed properties in epithelial cells. J Biol Chem 278(46): 45801–45810

106. Liu N, Gao F, Han Z, Xu X et al (2001) Hyaluronan synthase 3 overexpression promotes the growth of TSU prostate cancer cells. Cancer Res 61(13):5207–5214

107. Yu Q, Stamenkovic I (2000) Cell surface-localized matrix metalloproteinase-9 proteolytically activates TGF-beta and promotes tumor invasion and angiogenesis. Genes Dev 14(2):163–176

108. Tammi R, Saamanen AM, Maibach HI, Tammi M (1991) Degradation of newly synthesized high molecular mass hyaluronan in the epidermal and dermal compartments of human skin in organ culture. J Invest Dermatol 97(1):126–130

109. Turley E, Moore D (1984) Hyaluronate binding proteins also bind to fibronectin, laminin and collagen. Biochem Biophys Res Commun 121(3):808–814

110. Isemura M, Yosizawa Z, Koide T, Ono T (1982) Interaction of fibronectin and its proteolytic fragments with hyaluronic acid. J Biochem 91(2):731–734

111. Baccarani-Contri M, Vincenzi D, Cicchetti F, Mori G et al (1990) Immunocytochemical localization of proteoglycans within normal elastin fibers. Eur J Cell Biol 53(2):305–312

112. Tammi R, MacCallum D, Hascall VC, Pienimaki JP et al (1998) Hyaluronan bound to CD44 on keratinocytes is displaced by hyaluronan decasaccharides and not hexasaccharides. J Biol Chem 273(44):28878–28888

113. Tammi R, Tammi M (1991) Correlations between hyaluronan and epidermal proliferation as studied by [3H]glucosamine and [3H]thymidine incorporations and staining of hyaluronan on mitotic keratinocytes. Exp Cell Res 195(2):524–527

114. Oksala O, Salo T, Tammi R, Hakkinen L et al (1995) Expression of proteoglycans and hyaluronan during wound healing. J Histochem Cytochem 43(2):125–135

115. Foschi D, Castoldi L, Radaelli E, Abelli P et al (1990) Hyaluronic acid prevents oxygen free-radical damage to granulation tissue: a study in rats. Int J Tissue React 12(6):333–339

116. King SR, Hickerson WL, Proctor KG (1991) Beneficial actions of exogenous hyaluronic acid on wound healing. Surgery 109(1):76–84

117. Luke HJ, Prehm P (1999) Synthesis and shedding of hyaluronan from plasma membranes of human fibroblasts and metastatic and non-metastatic melanoma cells. Biochem J 343(Pt 1):71–75

118. Wisniewski HG, Naime D, Hua JC, Vilcek J et al (1996) TSG-6, a glycoprotein associated with arthritis, and its ligand hyaluronan exert opposite effects in a murine model of inflammation. Pflugers Arch 431(6 Suppl 2):R225–226

119. Kobayashi H, Terao T (1997) Hyaluronic acid-specific regulation of cytokines by human uterine fibroblasts. Am J Physiol 273(4 Pt 1):C1151–1159

120. Chen WY, Abatangelo G (1999) Functions of hyaluronan in wound repair. Wound Repair Regen 7(2):79–89

121. Frost SJ, Weigel PH (1990) Binding of hyaluronic acid to mammalian fibrinogens. Biochim Biophys Acta 1034(1):39–45

122. Weigel PH, Fuller GM, LeBoeuf RD (1986) A model for the role of hyaluronic acid and fibrin in the early events during the inflammatory response and wound healing. J Theor Biol 119(2):219–234

123. Knudson CB, Knudson W (1993) Hyaluronan-binding proteins in development, tissue homeostasis, and disease. FASEB J 7(13):1233–1241

124. Toole BP (2001) Hyaluronan in morphogenesis. Semin Cell Dev Biol 12(2):79–87

125. Jenkins RH, Thomas GJ, Williams JD, Steadman R (2004) Myofibroblastic differentiation leads to hyaluronan accumulation through reduced hyaluronan turnover. J Biol Chem 279(40):41453–41460

126. Desmouliere A, Guyot C, Gabbiani G (2004) The stroma reaction myofibroblast: a key player in the control of tumor cell behavior. Int J Dev Biol 48(5–6):509–517

127. Gabbiani G (2003) The myofibroblast in wound healing and fibrocontractive diseases. J Pathol 200(4):500–503

128. Meran S, Thomas D, Stephens P, Martin J et al (2007) Involvement of hyaluronan in regulation of fibroblast phenotype. J Biol Chem 282(35):25687–25697

129. Evanko SP, Angello JC, Wight TN (1999) Formation of hyaluronan- and versican-rich pericellular matrix is required for proliferation and migration of vascular smooth muscle cells. Arterioscler Thromb Vasc Biol 19(4):1004–1013

130. Knudson CB, Toole BP (1985) Fluorescent morphological probe for hyaluronate. J Cell Biol 100(5):1753–1758

131. Knudson CB, Knudson W (1990) Similar epithelial-stromal interactions in the regulation of hyaluronate production during limb morphogenesis and tumor invasion. Cancer Lett 52(2):113–122

132. Knudson CB, Munaim SI, Toole BP (1995) Ectodermal stimulation of the production of hyaluronan-dependent pericellular matrix by embryonic limb mesodermal cells. Dev Dyn 204(2):186–191

133. Webber J, Meran S, Steadman R, Phillips A (2009) Hyaluronan orchestrates transforming growth factor-beta1-dependent maintenance of myofibroblast phenotype. J Biol Chem 284(14):9083–9092

134. Clark RA, Lin F, Greiling D, An J et al (2004) Fibroblast invasive migration into fibronectin/fibrin gels requires a previously uncharacterized dermatan sulfate-CD44 proteoglycan. J Invest Dermatol 122(2):266–277

135. Acharya PS, Majumdar S, Jacob M, Hayden J et al (2008) Fibroblast migration is mediated by CD44-dependent TGF beta activation. J Cell Sci 121(Pt 9):1393–1402

136. Huebener P, Abou-Khamis T, Zymek P, Bujak M et al (2008) CD44 is critically involved in infarct healing by regulating the inflammatory and fibrotic response. J Immunol 180(4):2625–2633

137. Tolg C, Hamilton SR, Nakrieko KA, Kooshesh F et al (2006) Rhamm−/− fibroblasts are defective in CD44-mediated ERK1,2 motogenic signaling, leading to defective skin wound repair. J Cell Biol 175(6):1017–1028

138. Tammi RH, Tammi MI (2009) Hyaluronan accumulation in wounded epidermis: a mediator of keratinocyte activation. J Invest Dermatol 129(8):1858–1860

139. Sato H, Takahashi T, Ide H, Fukushima T et al (1988) Antioxidant activity of synovial fluid, hyaluronic acid, and two subcomponents of hyaluronic acid. Synovial fluid scavenging effect is enhanced in rheumatoid arthritis patients. Arthritis Rheum 31(1):63–71

140. Bates EJ, Harper GS, Lowther DA, Preston BN (1984) Effect of oxygen derived reactive species on cartilage proteoglycan-hyaluronate aggregates. Biochem Int 8(5):629–637

141. Deguine V, Menasche M, Ferrari P, Fraisse L et al (1998) Free radical depolymerization of hyaluronan by Maillard reaction products: role in liquefaction of aging vitreous. Int J Biol Macromol 22(1):17–22

142. Monzon ME, Fregien N, Schmid N, Falcon NS et al (2010) Reactive oxygen species and hyaluronidase 2 regulate airway epithelial hyaluronan fragmentation. J Biol Chem 285(34):26126–26134

143. Gao F, Koenitzer JR, Tobolewski JM, Jiang D et al (2008) Extracellular superoxide dismutase inhibits inflammation by preventing oxidative fragmentation of hyaluronan. J Biol Chem 283(10):6058–6066

144. Martin P (1997) Wound healing–aiming for perfect skin regeneration. Science 276(5309): 75–81

145. Beanes SR, Dang C, Soo C, Ting K (2003) Skin repair and scar formation: the central role of TGF-beta. Expert Rev Mol Med 5(8):1–22

146. Singer AJ, Clark RA (1999) Cutaneous wound healing. N Engl J Med 341(10):738–746

147. Lorenz HP, Whitby DJ, Longaker MT, Adzick NS (1993) Fetal wound healing. The ontogeny of scar formation in the non-human primate. Ann Surg 217(4):391–396

148. Longaker MT, Chiu ES, Adzick NS, Stern M et al (1991) Studies in fetal wound healing. V. A prolonged presence of hyaluronic acid characterizes fetal wound fluid. Ann Surg 213(4):292–296

149. Chen WY, Grant ME, Schor AM, Schor SL (1989) Differences between adult and foetal fibroblasts in the regulation of hyaluronate synthesis: correlation with migratory activity. J Cell Sci 94(Pt 3):577–584

150. David-Raoudi M, Tranchepain F, Deschrevel B, Vincent JC et al (2008) Differential effects of hyaluronan and its fragments on fibroblasts: relation to wound healing. Wound Repair Regen 16(2):274–287

151. Smith LT, Holbrook KA, Madri JA (1986) Collagen types I, III, and V in human embryonic and fetal skin. Am J Anat 175(4):507–521
152. Merkel JR, DiPaolo BR, Hallock GG, Rice DC (1988) Type I and type III collagen content of healing wounds in fetal and adult rats. Proc Soc Exp Biol Med 187(4):493–497
153. Volk SW, Wang Y, Mauldin EA, Liechty KW (2011) Diminished Type III Collagen Promotes Myofibroblast Differentiation and Increases Scar Deposition in Cutaneous Wound Healing. Cells Tissues Organs 194(1):25–37
154. Karvinen S, Pasonen-Seppanen S, Hyttinen JM, Pienimaki JP et al (2003) Keratinocyte growth factor stimulates migration and hyaluronan synthesis in the epidermis by activation of keratinocyte hyaluronan synthases 2 and 3. J Biol Chem 278(49):49495–49504
155. Pasonen-Seppanen SM, Maytin EV, Torronen KJ, Hyttinen JM et al (2008) All-trans retinoic acid-induced hyaluronan production and hyperplasia are partly mediated by EGFR signaling in epidermal keratinocytes. J Invest Dermatol 128(4):797–807
156. Rilla K, Pasonen-Seppanen S, Rieppo J, Tammi M et al (2004) The hyaluronan synthesis inhibitor 4-methylumbelliferone prevents keratinocyte activation and epidermal hyperproliferation induced by epidermal growth factor. J Invest Dermatol 123(4):708–714
157. Passi A, Sadeghi P, Kawamura H, Anand S et al (2004) Hyaluronan suppresses epidermal differentiation in organotypic cultures of rat keratinocytes. Exp Cell Res 296(2):123–134
158. Tammi R, Ripellino JA, Margolis RU, Maibach HI et al (1989) Hyaluronate accumulation in human epidermis treated with retinoic acid in skin organ culture. J Invest Dermatol 92(3):326–332
159. Agren UM, Tammi M, Tammi R (1995) Hydrocortisone regulation of hyaluronan metabolism in human skin organ culture. J Cell Physiol 164(2):240–248
160. Larjava H, Saarni H, Tammi M, Penttinen R et al (1980) Cortisol decreases the synthesis of hyaluronic acid by human aortic smooth muscle cells in culture. Atherosclerosis 35(2):135–143

Chapter 3
Lactoferrin Structure Function and Genetics

Abstract Lactoferrin is an iron-binding glycoprotein that belongs to the transferrin family. It consists of two globular domains, called N-lobe and C-lobe. Both lobes have the similar polypeptide-folding pattern and one metal-binding site. Its structure is conserved among species. As well as other members of transferrin family, lactoferrin exerts their biological functions by chelating free irons in body fluids in mammals. Lactoferrin acts as bacteriostatic agent, based on their ability to deprive iron essential for bacteria growth. Besides, its iron-chelating activity, lactoferrin positively or negatively regulates host immune response by controlling maturation, migration, cytokine secretion of innate and adaptive immune cells. Lactoferrin is detected in many body fluids, and contributes host defense through the bactericidal and immuno-modulating activities. The difference of the lactoferrin expression among the different species and tissue could be explained by the diversity of lactoferrin gene promoter region.

Keywords Bacteriostatic agent • Iron binding protein • Immune response

3.1 Introduction

Lactoferrin (lacto-transferrin) is a metal-binding glycoprotein, mainly secreted from glandular epithelial cells in the mammary gland. This molecule was first discovered as "red protein" in bovine milk and isolated from both human and bovine milk [1–3]. Lactoferrin has been considered as a functional factor for optimal dietary provision with iron and host defense. Based on the conserved three-dimensional (3D) structure and iron-chelating properties, lactoferrin belongs to the transferrin family along with ovotransferrin and melanotransferrin [4]. The members of the transferrin family exert their biological functions by regulating levels of free irons in body fluids in mammals, as they are involved in the iron transport in plasma and iron-uptake by intestinal mucosa. Furthermore, they act as bacteriostatic agent, based on their ability to deprive iron essential for bacteria growth [5]. In addition, lactoferrin potentially diminishes free radical damage by scavenging free iron at the site of inflammation [6]. Beside its ability as metal binding protein, lactoferrin binds to various types of mammalian cells, including neutrophils, macrophages (monocytes), natural killer (NK) cells, platelets, enterocytes, hepatocytes, dendritic cells, fibroblasts and osteoblasts [7–9], and regulates their migration, growth and differentiation. Lactoferrin is known as highly basic protein, and binds anionic molecules such as DNA, heparin, glycosaminoglycans, and endotoxins [10–12]. The multi-functionality of lactoferrin could be explained by diversity of target cells and molecules. Structural and functional data of lactoferrin suggests that its characteristic structure which contributes its diverse functions.

3.2 Lactoferrin Expression and Localization

Milk is known as abundant source of lactoferrin. Lactoferrin concentration changes during the lactation period. Its concentration in colostrum is much higher than that of milk. In human, its concentration is 1–2 mg/ml in mature milk and 10 mg/ml in colostrum [13]. In bovine, lactoferrin concentration is <0.2 mg/ml in mature milk and <1.5 mg/ml in colostrum [7, 14]. Lactoferrin expression could be observed only in mammals. The lactoferrin concentration in mature milk varies greatly among species. In human, primates, pig and mice, lactoferrin concentrations are relatively high (<2 mg/ml), whereas lactoferrin levels in rabbit, dog and rat are very little (<50 µg/ml) [15, 16]. Interestingly, species that have high lactoferrin concentration in their milk tend to have very little transferrin concentration in their milk and vice versa, suggesting that most mammals constantly contain transferrin or transferrin-like molecules in their milk [15]. In addition, lactoferrin could be detected in most exocrine secretions, including saliva, tears, semen and mucosal secretions [7].

Lactoferrin is also known as a major component of the secondary granules of polymorphonuclear neutrophils (PMNs) and is released in plasma during the infection or inflammation [17, 18]. In normal conditions, its concentration in plasma is as low as 0.4–2.0 µg/ml. It increases up to 200 µg/ml in microbial infections and autoimmune diseases [7, 19–21]. Plasma lactoferrin concentration is not correlated with the number of neutrophils. It is dependent on the extent of degranulation of the neutrophils [22]. Lactoferrin plasma levels change from the beginning of pregnancy. Lactoferrin concentration is progressively increased until 29th week, thereafter it remains high level until term [23].

3.3 Lactoferrin Structure

Lactoferrin consists of single polypeptide chain and glycan. Human and bovine lactoferrin consists of 692 and 689 amino acids, respectively [24, 25]. The amino acid sequence homology of human and bovine lactoferrin is about 69%.

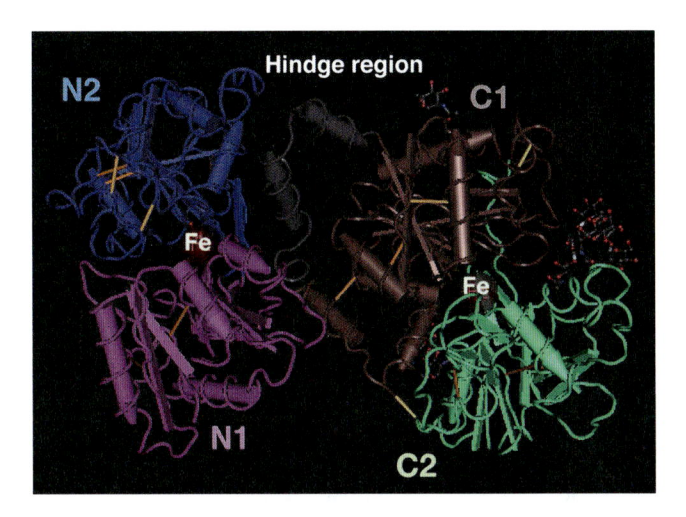

Fig. 3.1 Three dimensional structure of bovine lactoferrin. The N1 domain is colored *pink*, the N2 domain is *blue*, the hinge region is *gray*, the C1 domain is *brown*, and the C2 domain is *green*. Structures was drawn with the Cn3D 4.3 software program using PDB (Protein Data Bank) ID:1BLF (MMDB 55593) deposited by Moore et al. [28] in the Molecular Modeling Data base of NCBI (the National Center for Biotechnology Information, Bethesda, MD, USA)

The three-dimensional (3D) structure of human and bovine lactoferrin was identified [26–28]. Although both lactoferrin and transferrin have similar 3D structures, amino acid sequence homology between them is 60–65%. The difference was especially found in the surface-exposed sequence, which affects their physiochemical and biological properties. In contrast to transferrin, lactoferrin is a highly basic protein with an isoelectric point 8.7 [29, 30]. This basic nature of lactoferrin can explain its high affinity to many anionic molecules.

Lactoferrin consists of two globular domains, called N-lobe and C-lobe. Both lobes have the similar polypeptide-folding pattern and one metal binding site, whereas N-lobe has 40% sequence homology with C-lobe (Fig. 3.1). This has led to a theory of gene duplication, and subsequent formation of the two similar lobes. In human lactoferrin, N-lobe corresponds to amino acid residues 1–333 and C-lobe to 345–692. Both lobes consist of two sub-domains (N-lobe: N1 and N2 and C-lobe: C1 and C2). This two-lobe, four sub-domain structure of lactoferrin contributes its metal binding property. The structural feature of each lobe is that two subdomains enclose one metal ion along with one bicarbonate, which is prerequisite for metal ion binding, into the metal-binding cleft [31]. It consists of four protein ligands (Tyr92, Tyr 192, Asp60, His 253 in N-lobe of human lactoferrin), one metal ion, one bicarbonate ion. The protein ligands provide three negative charges to balance the three positive charge of the iron ion. Positive charge in arginine residues (Arg121 in N-lobe of human lactoferrin) in side-chain balances the one negative charge in bicarbonate anion. Metal ion is effectively sequestered away from the external environment. This structure can explain the tight and stable binding of lactoferrin to metal ion. It is difficult to remove the metal ion without destabilizing the whole protein structure by lowering pH [32]. This structure is highly conserved in the all member of the transferrin family.

The two lobes are linked by α-helix hinge region that is most sensitive part for protease cleavage. This inner lobe connecting peptide is helical in lactoferrin while it is irregular in transferrin.

As described above, lactoferrin is a highly basic protein. The cationic regions of lactoferrin are responsible for the binding of the negatively charged substrates, including lipopolysaccharide (LPS), CD14, heparin, lysozome, proteoglycans and DNA [33]. The degree of iron saturation does not affect the bindings [12]. The surface distribution of the positive charge is not ubiquitous. It is highly concentrated on the N-terminal (residues 1–7) and the outside region of the first helix (residues 13–30) of the N1 domain. The first N-terminal 5 residues (Gly-Arg-Arg-Arg-Arg) extend out of the protein surface, and their structure is flexible. In the first helix of the N1 domain, basic residues (Lys and Arg) are arrayed along the outside of the helix. They are major part of lactoferricin (LFcin), a strong anti-bacterial peptide derived from lactoferrin, and responsible for most of iron independent activities of lactoferrin [34]. In addition, positive charge is concentrated around the inner lobe connecting helix.

Despite the highly conserved primary structure of lactoferrin in different species [35, 36], diversity of post-translational modification could be observed in lactoferrin from different origin. The number and location of glycosylation sites vary in the different species [37]. Human lactoferrin contains three possible N-glycosylation sites, Asn138 in the N-lobe and Asn479 as well as Asn624 in the C-lobe [24]. They are glycosylated $in\ vivo$ about 94%, 100% and 9% of the molecules, respectively [38]. Bovine lactoferrin contains five N-glycosylation sites. Four sites (Asn233, 368, 476 and 545) are always utilized while the fifth (Asn281), located in the N-lobe, is glycosylated in about 30% of the molecules in bovine colostrum, but only in about 15% in mature milk. The sugars found in lactoferrin are N-acetylglucosamine, N-acetyllactosamine, N-acetylneuraminic acid galactose, fucose and mannose. The dominant glycoform is different among species. Glycans with polymannosidic structure are specific to bovine lactoferrin and not found on human lactoferrin. Lactoferrin secreted from neutrophils does not contain a fucose on the core, making it more similar to the glycan pattern of human serum transferrin [39]. The precise role of these glycans has not been established. Lactoferrin from different species showed identical 3D-structure, irrespective of the sites of glycosylation. Recombinant human lactoferrin produced in plants and microbes is glycosylated differently from original human lactoferrin. However, recombinant human lactoferrin expressed in $Aspergillus\ awamori$, termed as a talactoferrin, showed identical 3D structure compared with native human lactoferrin. The binding of human lactoferrin to iron and LPS is not affected by deglycosylation [40]. However, glycosylation at Asn281 protects bovine lactoferrin against cleavage by trypsin [41]. Overall, the attached glycan chains appear to have a little effect on lactoferrin structure and function.

3.4 Iron-Binding Properties of Lactoferrin

The most common metal ion associated with lactoferrin is iron ion (Fe^{3+}). One lactoferrin molecule can bind two iron ions with high affinity ($K \sim 10^{22}$ M) [42]. Bovine lactoferrin is only partially saturated with iron (>20%). Iron-deprived from of lactoferrin is called apo-lactoferrin, whereas iron-saturated form of lactoferrin is called holo-lactoferrin. Iron-binding to lactoferrin results in conformational change with increased protease resistance [31, 43, 44]. The structure of apo-lactoferrin is much less compact than that of holo-lactoferrin [45]. In addition, lactoferrin is capable of binding copper, manganese, zinc, gallium and aluminum ion with lower affinity [46].

Although human lactoferrin and transferrin have similar iron-binding sites, iron release from human lactoferrin begins at pH 4.0 and is complete at pH 2.5, whereas iron release from human transferrin begins at pH 6.0 and being complete at pH 4.0 [29]. The affinity of human and bovine lactoferrin for iron ion is about 260-fold, and 30-fold higher than those for transferrin, respectively [42]. This stability of lactoferrin is caused by cooperative interaction of the two lobes [31, 47]. The high affinity and stability of iron-binding by lactoferrin makes this protein not only a iron-carrier but also strong iron scavenger and antioxidant molecule.

As well as transferrin, supposed function of lactoferrin in the gastrointestinal tract is that promotion of iron absorption by enterocytes [48]. It has been proposed that both lactoferrin and iron are taken up by enterocytes [49, 50]. In support of this, specific lactoferrin receptor (intelectin/HL 1) has been identified in the brush border membrane of enterocytes [9, 51, 52]. Transfection of the human lactoferrin receptor gene in Caco-2 cells, an intestinal cell line, results in increased iron-uptake. However, lactoferrin knockout mouse shows increased iron level during sucking period. Iron delivery to neonate is not affected by the lactoferrin deficiency [53]. Indeed, major molecule for iron uptake in intestine is identified as divalent metal transporter 1 (DMT-1/DCT-1/NRAMP-2) [54–56]. These lines of observation suggest that the main function of lactoferrin as iron binding protein is not iron transporter but iron scavenger [57].

3.5 Lactoferrin Metabolism

Lactoferrin can avoid digestion in gastrointestinal tract. Lactoferrin (especially holo-lactoferrin) demonstrates remarkable resistance to proteolytic degradation by trypsin, chimo-trypsin and pepsin, especially in infants [43, 48]. Bovine lactoferrin is more resistant to proteolytic degradation than human lactoferrin. In adults, more than 60% of oral administrated bovine lactoferrin can survive in the stomach [58]. About 30% of iron in the human milk is bound to lactoferrin, and only >10% of milk lactoferrin is iron-saturated (holo-lactoferrin) [59]. It is important subject that whether oral administrated lactoferrin could be absorbed from the intestine and exerts its biological effect. In preterm human infants, neonatal pigs, and rats with colitis, bovine lactoferrin is detected in various body fluids after oral administration [60, 61]. Indeed, transport of oral administrated bovine lactoferrin into body fluid is not observed in adult rats [62, 63], suggesting that the barrier function of intestinal tract is not established in the neonatal pigs, and rats with colitis.

Most of plasma lactoferrin is derived from neutrophils. Secretion from neutrophil is induced by inflammation, infectious disease, during tumor development. Lactoferrin released in plasma is rapidly cleared from the circulation by liver [64, 65]. When bovine lactoferrin is injected intravenously, lactoferrin is removed within 7 h. As lactoferrin and its fragment could be detected in urine, the kidney seems to be involved in the lactoferrin clearance from circulation [66]. Alternatively, lactoferrin can be removed from circulation by receptor-mediated endocytosis into phagocytes such as monocytes and macrophages [67, 68].

Lactoferrin immunoreactivity is present in the pathologic brain lesions in several neurodegenerative disorders including Alzheimer's disease, Down syndrome, amyotrophic lateral sclerosis and Pick's disease [69, 70]. When bovine lactoferrin is injected intravenously, lactoferrin could cross the blood brain barrier (BBB) through receptor mediated endocytosis and transcytosis [71]. Interestingly, brain uptake of lactoferrin is much faster than that of transferrin [72]. *In vivo* diffusion of lactoferrin in brain extracellular space is regulated by interactions with heparan sulfate [73].

3.6 Anti-bacterial Activity of Lactoferrin

The broad distribution of lactoferrin in the gateways of the digestive, respiratory and reproductive system suggests that lactoferrin plays important roles in non-specific host defense in mammals. In fact, oral administrated lactoferrin has host-protecting effect against microbes [74].

The bactericidal activity of lactoferrin has been widely documented for both Gram-positive and Gram-negative bacteria [75]. It appears that two different mechanisms are involved in the activity. The first mechanism is based on its ability to sequester iron, which is essential for the growth of both Gram-positive and Gram-negative bacteria [5, 7, 75]. Lactoferrin is secreted from epithelial cells in apo-form, so that its iron-chelating activity can be directed against microbes. Due to the metabolic activity of bacteria, pH is decreased at infected and inflammatory tissues. As described above, lactoferrin can act as iron scavenger even at low pH conditions. This activity is dependent on the iron status of lactoferrin and is abrogated with their iron saturation [76].

In the second mechanism, lactoferrin exerts its anti-microbial effects by interacting target microbes directly. The mechanisms of function are different in Gram-negative and Gram-positive bacteria due to the differences in the bacterial membrane structure. In Gram-negative bacteria, treatment of lactoferrin destabilizes their outer membrane by removing LPS. LPS is a major structural component in the outer membrane of the Gram-negative bacteria, and is associated with their pathogenic capability. Lactoferrin binds to the lipid A of LPS with high affinity [10, 11]. The lactoferrin-mediated LPS release results in osmotic destruction of the outer membrane, and bacterial death [77, 78]. Porins, the major pore-forming proteins of the outer membrane of various Gram-negative bacteria, are participated in the process. Lactoferrin interacts with cell surface porins, thus weakening bonds between porins and LPS and inducing LPS release into the environment [79, 80].

The mechanism of lactoferrin action against Gram-positive bacteria is based on its direct binding to anionic molecules on the bacterial surface, such as lipoteichoic acid. This binding results in a reduction of negative charge on the bacterial cell wall and facilitates the accessibility of lysozyme to the underlying peptidoglycan, and increases the penetration of lysozyme into gram-negative bacteria [81].

In addition to lysozyme, lactoferrin cooperatively acts with other bacteriostatic agents present in exocrine secretions such as immunoglobulin, secretory leukocyte protease inhibitor, and anti-microbial peptides [77, 78, 82]. Lactoferrin also reinforces the activity of anti-microbial drugs such as lactam antibiotics and azole anti-fungal agents [83].

At the mucosal surface, lactoferrin contributes host defense by inhibiting biofilm formation by *Pseudomonas aeruginosa*. The biofilm formation with *P. aeruginosa* is observed in the patients of chronic lung infection. The proteolytic degradation of lactoferrin in the airway in the patients suggests the inhibitory effect of lactoferrin on the biofilm formation [84, 85]. Singh and his colleagues demonstrated that lactoferrin inhibits *P. aeruginosa* biofilm formation *in vitro* [86]. This activity is appeared to be dependent on the iron sequestration activity of lactoferrin. The lactoferrin concentration required to inhibit the biofilm formation (0.02 mg/ml) is fivefold less than that required to inhibits the growth of *P. aeruginosa*.

Lactoferrin degrades and inactivates proteins that are required for bacterial colonization by their proteolytic activity. This activity is inhibited by serine protease inhibitor. [57, 87].

Despite of its broad bacteriostatic activity, lactoferrin favors the growth of bacteria with low iron requirements such as lactic acid producing bacteria [88]. In addition, various bacterial species within *Neisseriaceae* and *Moxarella* have developed the mechanism for acquiring iron directly from lactoferrin/transferrin by using their own lactoferrin/transferrin receptors [89–91]. Lactoferrin (and transferrin) is ineffective for these species.

3.7 Bactericidal Activity of Lactoferrin Derived Peptides

The N-terminals of human and bovine lactoferrin contain cationic loop (N1 domain, residues 1–50). Lactoferricin is a peptide isolated following pepsin digestion of the N1 domain. Human lactoferricin (LFcin H) is corresponding to residues 1–47. Bovine lactoferricin (LFcin B) is corresponding to residues 17–41 [92–94]. The bactericidal effects of LFcin H and LFcin B are more prominent than that of human and bovine lactoferrin. They show the activity against a wide range of microorganisms including bacteria, yeast, fungi, viruses and protozoa [34, 92]. LFcin takes a β-sheet α-helix structure, and exerts their bactericidal activity by disruption or penetration of bacterial cell membrane. In addition, LFcin exhibits several types of biological function other than bactericidal activity, including tumor metastasis in mice [95], promotion of the cytokine synthesis and the phagocytic activity of PMNs [96, 97].

The bactericidal activity can be observed in several LFcin truncated peptides. LF11 is a peptide derived from the N-terminal loop of LFcin H (residues of 21–31). It is effective in neutralizing bacterial LPS [98, 99]. The anti-bacterial center of LFcin B is identified as RRWQWR [94]. Another known antimicrobial peptide is lactoferrampin (residues of 265–284). This α-helix containing peptide has been identified in the N1-domain of bovine lactoferrin [100]. This peptide displayed antimicrobial activity against bacteria *E. coli, Bacillus subtilis, P. aeruginosa* and yeast.

3.8 Anti-viral Activity of Lactoferrin

Lactoferrin processes antiviral activity against a broad range of DNA and RNA virus that infect human and animals. Lactoferrin blocks the internalization of adenovirus, poliovirus type 1, hepatitis C virus (HCV), human immunodeficiency virus (HIV), herpes simplex virus (HSV) type 1 and cytomegalovirus (CMV) [101–104]. Lactoferrin inhibits the replication of rotavirus, HCV, HIV and HSV type 1 and 2 [102, 105–107]. The mechanism of anti-viral activity of lactoferrin is not fully characterized. The ability of lactoferrin to bind glycosaminoglycan could explain the mechanism [108]. One of the most important receptor for virus infection is proteoglycan, especially heparan sulfate. Lactoferrin can bind to cell surface heparan sulfate, and prevents the virus contact to the host cells [109]. Lactoferrin inhibits HIV entry into cells by binding V3 loop of the gp120 HIV envelope protein [110]. Lactoferrin can prevent HCV adsorption to target cells by binding to the envelope proteins of E1 and E2, the docking protein in the viral coat of the HCV [111, 112].

3.9 Anti-parasitic and Anti-fungal Activity of Lactoferrin

In many cases, lactoferrin acts against parasites by breaching their membrane integrity. Among fungal pathogens responsible for opportunistic infections, species of the genus *Candida* have a central contribution. Lactoferrin secreted in mucosal surface has antifungal activity against *Candida* due to membrane perturbation [83]. Intestinal amoebiasis is caused by infection of *Entamoeba histolytica*. Lactoferrin kills *E. histolytica* by binding the lipids of their membrane, and disrupts the membrane structure [113]. On the other hand, lactoferrin induces tolerance to infection of *Toxoplasma gondii* by inhibiting the intracellular growth of *T. gondii* [114]. The anti-parasitic activity of lactoferrin against *Pneumocystis carinii* is dependent on its iron-chelating activity [115]. Lactoferrin inhibits accumulation of infectious isoform of prion protein (PrP^{sc}) in scrapie-infected cells [116]. Lactoferrin can interact with both PrP^{sc} and its cellular isoform (PrP^c), and inhibits the uptake of PrP^c.

3.10 Immunoregulatory Properties of Lactoferrin

Lactoferrin is constitutively detected in various body fluids in normal physiological conditions. In response to tissue infection or inflammation, both innate leukocytes (macrophages, dendritic cells and natural killer (NK) cells) and adaptive immune cells (T-cells and B-cells) are exposed to high concentration of lactoferrin as a result of degranulation of neutrophils. Accumulating evidences suggest that lactoferrin can control host immune response by regulating their maturation, migration, and secretion of cytokines and chemokines [117, 118].

The immunoregulatory function of lactoferrin had been suggested since 1980. In a patient suffering from recurrent infections, lactoferrin expression in neutrophil is deficient whereas its expression in glandular secretions is normal [119]. Accordingly, the results of animal study suggest the regulatory role of lactoferrin in immune response. Transgenic mice expressing human lactoferrin exhibit enhanced T helper type 1 (Th1) response to *Staphylococcus aureus* [120]. Oral administration of lactoferrin diminished the susceptibility to tuberculosis in β2-microglobulin knockout mice [121]. Mixture of lactoferrin with monophosphoryl lipid A is an efficient adjuvant of the humoral and cellular immune response [122].

Phagocytosis is a central component of the innate immune response. Professional phagocytes (neutrophils, macrophages, and dendritic cells) recognize and engulf foreign extracellular material. The effect of lactoferrin on their phagocytic activity is well characterized. Lactoferrin expression is observed even in resting neutrophils and involved in its binding to microbes [123]. Lactoferrin directly interacts with neutrophils and promotes their phagocytic activity and reactive oxygen species (ROS) production [96, 124]. The molecular mechanism of the effects of lactoferrin on neutrophil phagocytosis is not clear. Complement factor C3 is required for phagocytic activity of neutrophils. However, the effect of lactoferrin on the classical and alternative pathways of complement is not clear. Both positive and negative effects of lactoferrin on complement pathway have been reported [8]. Indeed, the effect of lactoferrin on neutrophil migration is still controversial. Lactoferrin antagonizes neutrophil migration towards IL-8, C5a and fMLP, suggesting that lactoferrin inhibits excessive neutrophil activation and inflammation leading to host tissue damage [125]. However, some other reports have demonstrated that lactoferrin promotes the neutrophil chemotaxis [124, 126].

Monocyte is the first mammalian cell whose lactoferrin receptor is identified [127]. Lactoferrin stimulates phagocytosis of monocytes, and promotes their migration and ROS production [126]. Lactoferrin increases CD40 expression and IL-6 production in THP-1 cells (human acute monocytic leukemia cell line) and promotes the differentiation of THP-1 cells to macrophages [128].

Macrophages are professional phagocytes involved in innate immune response through phagocytosis of infected microbes. Lactoferrin can interact with macrophages and regulates their phagocytic activity. Accordingly, oral administration of lactoferrin in mice increases phagocytic activity of macrophages injected with inactivated *Candida albicans* [129]. Lactoferrin promotes phagocytosis of macrophages and promotes IL-8, TNF-α and nitric oxide (NO) production *in vitro* [130].

Lactoferrin stimulates the cytotoxic activity of natural NK cells [131, 132]. Oral administration of bovine lactoferrin increases NK cell activity in mice by increasing the production of IL-18 and type-I IFN in the small intestine [133]. Likewise, oral administration of lactoferrin increases NK cells and CD4+ CD8+ T cell in the lamina propria of small intestine [134] or in blood and lymphoid tissues [135].

In addition to innate immune response, lactoferrin can regulate antigen specific adaptive immune response. Macrophages are involved in the adaptive immune system by presenting antigen to CD4+ T-lymphocytes. Stimulation of antigen specific CD4+ T-cells requires presentation of antigen peptides by macrophages via MHC II and co-stimulatory cell surface molecules such as CD80, CD86 and CD40. Lactoferrin can regulate the levels of the co-stimulatory molecules on the surface of macrophages [117, 118]. IFNγ-induced MHC II expression is decreased in macrophages infected with BCG (Bacillus Calmette-Guérin) [136, 137]. Lactoferrin reverses BCG-mediated suppression of MHC II expression in macrophages [138, 139]. Likewise, lactoferrin promotes CD40 expression in macrophages [140]. The promoting effect of bovine lactoferrin on CD40 expression is TLR4-dependent whereas the effect of lactoferrin on IL-6 secretion is TLR4-independent [140].

Lactoferrin receptor expression can be detected in CD4+, CD8+, and γδ T-cells [141]. Lactoferrin promotes the maturation of T-cells towards CD4+ cells [142, 143]. Oral administration of lactoferrin in mice strongly elevates the pool of CD3+ T-cells and CD4+ T-cells [144] and γδ T-cells [145]. Lactoferrin increases CD4 expression in Jurkat T-cells line [143]. Orally administrated lactoferrin reconstitutes a T-cell mediated immune response by renewal of the T-cell pool in cyclophsph-amide (CP)-treated mice [146]. Lactoferrin promotes differentiation of naïve T-cells to either Th1 or Th2 depending on antigen. Intraperitoneal injection of lactoferrin increases the macrophage production of IL-12 [147]. IL-12 acts directly to CD4+ T-cells and promotes IFN-γ production, and induces Th1 cytokine dominant environment. Many studies indicate that lactoferrin can regulate the maturation and differentiation of T-cells and Th1/Th2 cytokine balance [148]. As described, lactoferrin transgenic mice exhibit enhanced Th1 response to *S. aureus* [120]. Oral administration of bovine lactoferrin in mice induces production of IFN-γ and IL-12 in response to HSV type 1 infection [149]. According to the results of quantitative RT-PCR analysis, oral administration of bovine lactoferrin in mice increases IFN-β and IL-12p40 mRNA expression in small intestine [145]. Oral administration of bovine lactoferrin in patients with chronic hepatitis C increases IFN-γ positive T-cells in the peripheral blood, along with the elevation of serum IL-18 level [150]. Bovine lactoferrin also increases synthesis of IL-18 in small intestinal epithelium [134, 151]. IL-18 enhances Th1 type T-cell response and generates CD8+ T-cells [152]. Conversely, lactoferrin decreases IFN-γ and increases IL-10 in an infection model of *Toxoplasma gondii*, suggesting that promotion of a Th2 response [148]. The effects of lactoferrin on the newly defined T-cell subsets, regulatory T cells (Treg) and Th17 cells, are currently unknown.

Dendritic cells act as professional antigen-presenting cells in the adaptive immune system because of their ability to induce a primary immune response by activation of naive T-cells. Lactoferrin acts as a maturation factor for human dendritic cells [126, 153]. Recombinant human lactoferrin enhances the expression of human leukocyte antigen (HLA) class II, CD83, CD80, and CD86 and CXCR4 and CCR7 chemokine receptors. Dendritic cells matured by lactoferrin display an enhanced release of IL-8 and CXCL10 and a significantly reduced production of IL-6, IL-10, and CCL20. Culturing BCG-infected dendritic cells with lactoferrin also enhanced their ability to respond to IFN-γ activation through up-regulation of maturation markers: MHC I, MHC II and the ratio of CD86:CD80 surface expression [154].

Lactoferrin can bind B-cells and regulates their response [155]. The antibody isotype of a B-cell changes during B-cell maturation. B-cells from spleen of newborn mice express mainly surface IgM, followed by expression of IgD and later by formation of complement receptors (C3R). Lactoferrin promotes maturation of B-cells from newborn mice by inducing changes in the ratio of IgM/IgD-bearing cells and in the appearance of complement C3 receptors [156]. The same study demonstrates that incubation of immature B-cells with lactoferrin enhances their ability to promote proliferation of antigen-specific T-cells [156]. Some daughter cells of the activated B-cells undergo class switching, a biological mechanism that changes a B cell's production of antibody from IgM/IgD to another (IgE, IgA or IgG), that have defined roles in the immune system. Oral administration of lactoferrin increases secreation of IgA and IgG from murine Peyer's patches [157] and instestinal secreations [144]. Moreover, lactoferrin can prevent methotrexate (MXT)-induced suppression of immune response to sheep erythrocytes in mice [158]. The effect of lactoferrin depends on the terminal *N*-acetylneuraminic acid [159]. These lines of observations indicate that lactoferrin regulates B-cell function as APCs to regulates T-cell response.

3.11 Anti-inflammatory Activity of Lactoferrin

The protective role of lactoferrin against inflammation is partially depends on its ability to bind free iron ion and bacterial endotoxin such as LPS. Animal studies have shown that lactoferrin administration diminishes inflammation induced by challenge with bacteria or LPS. Lactoferrin protects gut mucosal integrity during endotoxemia induced by LPS in mice [160]. Intravenous administration of lactoferrin peptide prior to LPS injection protects mice against endotoxin lethality [161].

When tissues are infected with microbes, activated granulocytes release pro-inflammatory cytokines, such as interleukin-1β (IL-1β), IL-6, Il-8, granulocyte-macrophage colony-stimulating factor (GM-CSF), and tumor necrosis factor

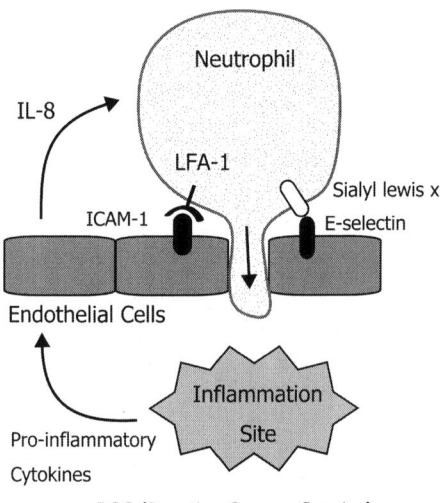

Fig. 3.2 The role of neutrophils in the inflammatory response. Pro-inflammatory cytokines induce the expression of IL-8 in endothelial cells. IL-8 is a potent chemoattractant and activator of neutrophils. Thes pro-inflammatory cytokines also increase the expression of E-selectin (CD62E), ICAM-1 (CD54) in endothelial cells. E-selectin and ICAM-1 interacts with sialylated Lewis X (SLex) and LFA-1 (αLβ2 integrin, CD11a/CD18) on the surface of neutrophils and mediates neutrophil attachment to the endothelial cells. The activated neutrophils interact with the endothelial cells and migrate into the inflammatory tissues

α (TNF-α) into plasma. These pro-inflammatory cytokines increase the expression of E-selectin (CD62E), ICAM-1 (CD54) in endothelial cells. E-selectin and ICAM-1 interacts with sialylated Lewis X (SLex) and LFA-1 (αLβ2 integrin, CD11a/CD18) on the surface of neutrophils and are involved in neutrophil attachment to the endothelial cells. On the other hand, the pro-inflammatory cytokines increase vascular permeability and accelerate IL-8 expression. IL-8 is a strong chemotaxis factor for neutrophils, and activates LFA-1 expression in neutrophils. Consequently, the activated neutrophils are attached to vascular endothelial cells and infiltrate into the inflammatory tissues (Fig. 3.2). The neutrophils gather at the inflammatory tissue secrete second granules. Since lactoferrin is known as a major component of the neutrophil secondary granules, lactoferrin concentration is dramatically increased at the sites of inflammation [8]. Accumulating evidences indicate that elevation of lactoferrin levels in some chronic inflammatory diseases, such as inflammatory bowel disease, gastrointestinal disorders, allergenic skin reactions and arthritis [162–166].

The results of *in vitro* and *in vivo* experiments suggest that the protective effect of lactoferrin on inflammation is partially due to inhibition of the pro-inflammatory cytokine production. Lactoferrin antagonizes production of TNF-α, IL-1β and IL-6 in mononuclear cells in response to LPS or BCG infection [167–171]. Pegtylated

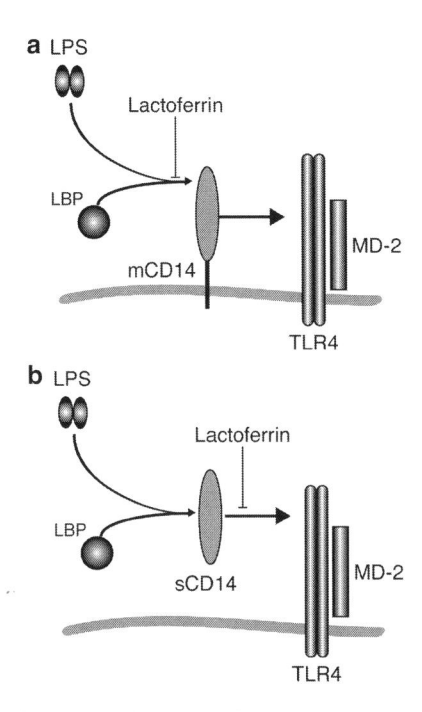

Fig. 3.3 The role of lactoferrin on LPS-mediated TLR4 activation. (**a**) LPS signal is transmitted through CD14/TLR4 to the intracellular mediators. Membrane-bound CD14 (mCD14) is expressed mainly on circulating monocytes and tissue macrophages. LPS (lipopolysaccharide) is bound by LBP (LBS-binding protein). The LPS/LBP complex binds to the mCD14. The binding of LPS-LBP to mCD14 recruits receptor complex which consists of TLR4 (toll like receptor 4) and MD-2. Lactoferrin inhibits the binding of LPS-LBP complex to mCD14. (**b**) In endothelial cells and epithelial cells, sCD14 (a soluble form of CD14) facilitates LPS/LBP binding to TLR4. Lactoferrin antagonizes sCD14 binding to TLR4/MD-2 complex

lactoferrin inhibits production of pro-inflammatory cytokines (TNF-α and IL-6) and hemorrhagic changes in response to liver injury induced by D-galactosamine and LPS in rats [172]. In endothelial cell, lactoferrin antagonizes the promoting effect of LPS on E-selectin, ICAM-1 and IL-8 expression [173, 174].

Additionally, the immunomodulatory properties of lactoferrin could be explained by the ability of lactoferrin to interact with proteoglycans. Lactoferrin competes with IL-8 for binding to cell surface proteoglycans, and attenuates its chemokine activity for neutrophils [174].

Overall, LPS-binding property of lactoferrin plays an important role in the immunomodulatory activity (Fig. 3.3). LPS is a potent initiator of an inflammatory response and serves as an indicator of bacterial infection. LPS interacts with LPS-binding protein (LBP) and CD14 [175]. CD14 exists in two forms. mCD14 is a membrane-bound glycosylphosphatidylinositol (GPI)-anchored glycosylated protein. sCD14 is a soluble form of CD14 which can be detected in the circulation. In monocytes and macrophages, mCD14 acts as LPS receptor. LPS forms a complex with LBP that, in turn, binds to the mCD14 on the surface of monocytes or macrophages. mCD14 enables LPS to be transferred to the LPS receptor complex, which is composed of Toll-like receptor 4 (TLR4) and MD-2 and induces the release of the pro-inflammatory cytokines [176]. As described above, lactoferrin interacts with lipid A part of LPS with high affinity, and inhibits bindings of LPS/LBP complex to mCD14, and antagonizes the production of pro-inflammatory cytokines [10, 11]. Endothelial and epithelial cells do not express mCD14. Instead, sCD14 is involved in TLR4-dependent recognition of LPS. Lactoferrin can interact with sCD14 in serum and inhibits endothelial cell activation induced by the sCD14-LPS complex [173, 176].

In contrast, oral administration of lactoferrin in rats with colitis increases IL-4 and IL-10 expression [61]. They are known as anti-inflammatory cytokine. Generally, lactoferrin released from neutrophils seems to act as a negative feedback regulator in inflammatory response by decreasing the production of pro-inflammatory cytokines or by neutralizing the effect of bacterial endotoxin such as LPS.

3.12   Anti-oxidative Effect of Lactoferrin

In infected and inflammatory tissue, ROS are abundantly produced by free iron released from necrosed tissue. Production of excessive free ion results in ROS overproduction and contributes to the pathogenesis of septic shock. In addition, activated granulocytes produce ROS to kill infected microbes. Lactoferrin acts as an iron scavenger, and diminishes the oxidative damage to tissues by inhibiting the ROS production. Lactoferrin also suppresses LPS-induced neutrophil production of hydrogen peroxide [173]. On the other hand, in acidic environments such as in the phagosomes, lactoferrin acts as an iron provider for the catalysis of ROS production and increases bactericidal activity of neutrophils [124, 177].

3.13   Genetics

The lactoferrin gene was identified at the chromosome level in a set of different species (human chromosome 3p21.3 [178, 179] and mouse chromosome 9 [180]) and its size ranges from 23 to 35 kb [181]. Lactoferrin gene consists of 17 exons. The number of amino acids in each exon and exon-intron junction is highly conserved among species. However, the size of introns varies widely among species [181]. In human lactoferrin, the sequence of first two exons is similar to that of human transferrin, suggesting that both lactoferrin and transferrin are derived from a common ancient iron binding protein [180].

Lactoferrin gene was originally discovered as an estrogen responsible gene [182]. In the mouse uterus, lactoferrin mRNA level is dramatically increased (>300-fold) in response to estrogen (17β-estradiol). Estrogen receptor α (ERα) binds estrogen response element (ERE) that locates at neighbor of lactoferrin promoter region. ERE acts as an enhancer in response to estrogen stimulation, and positively regulates the lactoferrin gene expression [35]. On the other hand, mouse and human ERE are overlapped with chicken ovalbumin up-stream promoter (COUP) [183]. In mice, COUP-TF (COUP transcription factor) acts as a competitive repressor for ER-mediated activation of the mouse lactoferrin gene. The ratio of ER/COUP-TF in the transfected cells appears to be critical for estrogen-stimulated lactoferrin gene promoter activity [183]. In addition to ERE, human lactoferrin gene contains an extended estrogen response element half site (ERRE) [184]. This region increases the sensitivity of human lactoferrin gene to estrogen. In mouse uterus, a repressor of estrogen receptor activity (REA) is identified as a negative regulator for estrogen-enhanced lactoferrin mRNA expression. This diversity of the transcription factor can account for difference of the lactoferrin gene expression among the different species and tissue.

Retinoic acid stimulates the lactoferrin expression in the embryonic stem cells [185]. Mouse lactoferrin ERE can function as a retinoic acid response element [186]. Peroxisome proliferator-activated receptor-γ (PPARγ) negatively regulates lactoferrin mRNA expression in mouse liver [187]. SP1 (specific protein 1), C/EBPε, CDP/cut, and Ets are transcription factors involved in the regulation of lactoferrin gene expression in mouse and human. C/EBP is a positive regulator of lactoferrin gene expression in neutrophils [188]. CDP/cut is a negative regulator of lactoferrin gene expression. Interestingly, the phenotypes of C/EBPε knockout mice are similar to patient with neutrophils specific secondary granule deficiency [189].

The cDNA sequence of human mammary lactoferrin 99.7% is agreement with cDNA of neutrophil lactoferrin. Amino acid homology between them is 97%. In bovine, the mRNA encodes 708 amino acids with 19 signal peptides. Regulation of lactoferrin synthesis is tissue specific. In mammary gland, lactoferrin expression is regulated by prolactin, and unaffected by estrogen [190], whereas estrogen regulation of lactoferrin expression is observed in uterus and vaginal epithelium [182]. The synthesis of lactoferrin in endometrium is regulated by both estrogen and epidermal growth factor (EGF) [191]. The expression of lactoferrin is controlled during the stage of embryogenesis [192]. It is limited to the preimplantation embryo, postimplantation neutrophils, and epithelial cells of the developing digestive and respiratory tract.

In addition to full-length lactoferrin, a truncated form of lactoferrin (delta lactoferrin) is produced by alternative splicing. Delta lactoferrin is a cytoplasmic protein and acts as a transcription factor that promotes DcpS, Skp1 and Bax genes provoking cell cycle arrest and apoptosis [193, 194].

3.14 Other Functions

Lactoferrin is the milk protein with the highest levels of amylase, RNase and ATPase activity [195]. The RNase activity is detected only in the two isoforms of lactoferrin (lactoferrin β and γ). The basis of the enzymatic activity is not understood.

References

1. Johanson B (1960) Isolation of an iron-containing red protein from human milk. Acta Chem Scand 14(2):510–512
2. Montreuil J, Tonnelat J, Mullet S (1960) Preparation Et Proprietes De La Lactosiderophiline (Lactotransferrine) Du Lait De Femme. Biochim Biophys Acta 45(3):413–421
3. Groves ML (1960) The isolation of a red protein from milk. J Am Chem Soc 82(13):3345–3350
4. Baker HM, Anderson BF, Baker EN (2003) Dealing with iron: common structural principles in proteins that transport iron and heme. Proc Natl Acad Sci USA 100(7):3579–3583
5. Bullen JJ, Rogers HJ, Leigh L (1972) Iron-binding proteins in milk and resistance to Escherichia coli infection in infants. Br Med J 1(5792):69–75
6. Nuijens JH, van Berkel PH, Schanbacher FL (1996) Structure and biological actions of lactoferrin. J Mammary Gland Biol Neoplasia 1(3):285–295
7. Levay PF, Viljoen M (1995) Lactoferrin: a general review. Haematologica 80(3):252–267
8. Legrand D, Elass E, Carpentier M, Mazurier J (2005) Lactoferrin: a modulator of immune and inflammatory responses. Cell Mol Life Sci 62(22):2549–2559
9. Suzuki YA, Lopez V, Lonnerdal B (2005) Mammalian lactoferrin receptors: structure and function. Cell Mol Life Sci 62(22):2560–2575
10. Appelmelk BJ, An YQ, Geerts M, Thijs BG et al (1994) Lactoferrin is a lipid a-binding protein. Infect Immun 62(6):2628–2632
11. Brandenburg K, Jurgens G, Muller M, Fukuoka S et al (2001) Biophysical characterization of lipopolysaccharide and lipid A inactivation by lactoferrin. Biol Chem 382(8):1215–1225
12. Mann DM, Romm E, Migliorini M (1994) Delineation of the glycosaminoglycan-binding site in the human inflammatory response protein lactoferrin. J Biol Chem 269(38):23661–23667
13. Houghton MR, Gracey M, Burke V, Bottrell C et al (1985) Breast milk lactoferrin levels in relation to maternal nutritional status. J Pediatr Gastroenterol Nutr 4(2):230–233
14. Steijns JM, van Hooijdonk AC (2000) Occurrence, structure, biochemical properties and technological characteristics of lactoferrin. Br J Nutr 84(Suppl 1):S11–S17
15. Masson PL, Heremans JF (1971) Lactoferrin in milk from different species. Comp Biochem Physiol B 39(1):119–129
16. Hoshino A, Hisayasu S, Shimada T (1996) Complete sequence analysis of rat transferrin and expression of transferrin but not lactoferrin in the digestive glands. Comp Biochem Physiol B Biochem Mol Biol 113(3):491–497
17. Masson PL, Heremans JF, Schonne E (1969) Lactoferrin, an iron-binding protein in neutrophilic leukocytes. J Exp Med 130(3):643–658
18. Birgens HS (1985) Lactoferrin in plasma measured by an ELISA technique: evidence that plasma lactoferrin is an indicator of neutrophil turnover and bone marrow activity in acute leukaemia. Scand J Haematol 34(4):326–331
19. Lash JA, Coates TD, Lafuze J, Baehner RL et al (1983) Plasma lactoferrin reflects granulocyte activation in vivo. Blood 61(5):885–888
20. Nuijens JH, Abbink JJ, Wachtfogel YT, Colman RW et al (1992) Plasma elastase alpha 1-antitrypsin and lactoferrin in sepsis: evidence for neutrophils as mediators in fatal sepsis. J Lab Clin Med 119(2):159–168
21. Gutteberg TJ, Rokke O, Andersen O, Jorgensen T (1989) Early fall of circulating iron and rapid rise of lactoferrin in septicemia and endotoxemia: an early defence mechanism. Scand J Infect Dis 21(6):709–715
22. Masson PL, Heremans JF, Ferin J (1968) Presence of an iron-binding protein (lactoferrin) in the genital tract of the human female. I. Its immunohistochemical localization in the endometrium. Fertil Steril 19(5):679–689
23. Sykes JA, Thomas MJ, Goldie DJ, Turner GM (1982) Plasma lactoferrin levels in pregnancy and cystic fibrosis. Clin Chim Acta 122(3):385–393

24. Rey MW, Woloshuk SL, deBoer HA, Pieper FR (1990) Complete nucleotide sequence of human mammary gland lactoferrin. Nucleic Acids Res 18(17):5288
25. Pierce A, Colavizza D, Benaissa M, Maes P et al (1991) Molecular cloning and sequence analysis of bovine lactotransferrin. Eur J Biochem 196(1):177–184
26. Anderson BF, Baker HM, Dodson EJ, Norris GE (1987) Structure of human lactoferrin at 3.2-a resolution. Proc Natl Acad Sci USA 84(7):1769–1773
27. Anderson BF, Baker HM, Norris GE, Rice DW (1989) Structure of human lactoferrin – crystallographic structure-analysis and refinement at 2.8-a resolution. J Mol Biol 209(4):711–734
28. Moore SA, Anderson BF, Groom CR, Haridas M (1997) Three-dimensional structure of diferric bovine lactoferrin at 2.8 A resolution. J Mol Biol 274(2):222–236
29. Moguilevsky N, Retegui LA, Masson PL (1985) Comparison of human lactoferrins from milk and neutrophilic leucocytes. Relative molecular mass, isoelectric point, iron-binding properties and uptake by the liver. Biochem J 229(2):353–359
30. Furmanski P, Li ZP, Fortuna MB, Swamy CV et al (1989) Multiple molecular forms of human lactoferrin. Identification of a class of lactoferrins that possess ribonuclease activity and lack iron-binding capacity. J Exp Med 170(2):415–429
31. Anderson BF, Baker HM, Norris GE, Rumball SV et al (1990) Apolactoferrin structure demonstrates ligand-induced conformational change in transferrins. Nature 344(6268):784–787
32. Mazurier J, Spik G (1980) Comparative study of the iron-binding properties of human transferrins. I. Complete and sequential iron saturation and desaturation of the lactotransferrin. Biochim Biophys Acta 629(2):399–408
33. van Berkel PH, Geerts ME, van Veen HA, Mericskay M et al (1997) N-terminal stretch Arg2, Arg3, Arg4 and Arg5 of human lactoferrin is essential for binding to heparin, bacterial lipopolysaccharide, human lysozyme and DNA. Biochem J 328(Pt 1):145–151
34. Wakabayashi H, Takase M, Tomita M (2003) Lactoferricin derived from milk protein lactoferrin. Curr Pharm Des 9(16):1277–1287
35. Teng CT (2002) Lactoferrin gene expression and regulation: an overview. Biochem Cell Biol 80(1):7–16
36. Teng CT (2010) Lactoferrin: the path from protein to gene. Biometals 23(3):359–364
37. Spik G, Coddeville B, Mazurier J, Bourne Y et al (1994) Primary and three-dimensional structure of lactotransferrin (lactoferrin) glycans. Adv Exp Med Biol 357:21–32
38. van Berkel PH, van Veen HA, Geerts ME, de Boer HA et al (1996) Heterogeneity in utilization of N-glycosylation sites Asn624 and Asn138 in human lactoferrin: a study with glycosylation-site mutants. Biochem J 319(Pt 1):117–122
39. Derisbourg P, Wieruszeski JM, Montreuil J, Spik G (1990) Primary structure of glycans isolated from human leucocyte lactotransferrin. Absence of fucose residues questions the proposed mechanism of hyposideraemia. Biochem J 269(3):821–825
40. van Berkel PH, Geerts ME, van Veen HA, Kooiman PM et al (1995) Glycosylated and unglycosylated human lactoferrins both bind iron and show identical affinities towards human lysozyme and bacterial lipopolysaccharide, but differ in their susceptibilities towards tryptic proteolysis. Biochem J 312(Pt 1):107–114
41. van Veen HA, Geerts ME, van Berkel PH, Nuijens JH (2004) The role of N-linked glycosylation in the protection of human and bovine lactoferrin against tryptic proteolysis. Eur J Biochem 271(4):678–684
42. Aisen P, Liebman A (1972) Lactoferrin and transferrin – comparative study. Biochim Biophys Acta 257(2):314–323
43. Brock JH, Arzabe F, Lampreave F, Pineiro A (1976) The effect of trypsin on bovine transferrin and lactoferrin. Biochim Biophys Acta 446(1):214–225
44. Brines RD, Brock JH (1983) The effect of trypsin and chymotrypsin on the in vitro antimicrobial and iron-binding properties of lactoferrin in human milk and bovine colostrum. Unusual resistance of human apolactoferrin to proteolytic digestion. Biochim Biophys Acta 759(3):229–235

45. Grossmann JG, Neu M, Pantos E, Schwab FJ et al (1992) X-ray solution scattering reveals conformational changes upon iron uptake in lactoferrin, serum and ovo-transferrins. J Mol Biol 225(3):811–819
46. Harrington JP, Stuart J, Jones A (1987) Unfolding of iron and copper complexes of human lactoferrin and transferrin. Int J Biochem 19(10):1001–1008
47. Ward PP, Zhou X, Conneely OM (1996) Cooperative interactions between the amino- and carboxyl-terminal lobes contribute to the unique iron-binding stability of lactoferrin. J Biol Chem 271(22):12790–12794
48. Lonnerdal B, Iyer S (1995) Lactoferrin: molecular structure and biological function. Annu Rev Nutr 15:93–110
49. Mikogami T, Heyman M, Spik G, Desjeux JF (1994) Apical-to-basolateral transepithelial transport of human lactoferrin in the intestinal cell line HT-29cl.19A. Am J Physiol 267 (2 Pt 1):G308–G315
50. Gislason J, Douglas GC, Hutchens TW, Lonnerdal B (1995) Receptor-mediated binding of milk lactoferrin to nursing piglet enterocytes: a model for studies on absorption of lactoferrin-bound iron. J Pediatr Gastroenterol Nutr 21(1):37–43
51. Suzuki YA, Shin K, Lonnerdal B (2001) Molecular cloning and functional expression of a human intestinal lactoferrin receptor. Biochemistry 40(51):15771–15779
52. Tsuji S, Uehori J, Matsumoto M, Suzuki Y et al (2001) Human intelectin is a novel soluble lectin that recognizes galactofuranose in carbohydrate chains of bacterial cell wall. J Biol Chem 276(26):23456–23463
53. Ward PP, Mendoza-Meneses M, Cunningham GA, Conneely OM (2003) Iron status in mice carrying a targeted disruption of lactoferrin. Mol Cell Biol 23(1):178–185
54. Andrews NC (2000) Iron homeostasis: insights from genetics and animal models. Nat Rev Genet 1(3):208–217
55. Fleming MD, Trenor CC 3rd, Su MA, Foernzler D et al (1997) Microcytic anaemia mice have a mutation in Nramp2, a candidate iron transporter gene. Nat Genet 16(4):383–386
56. Gunshin H, Mackenzie B, Berger UV, Gunshin Y et al (1997) Cloning and characterization of a mammalian proton-coupled metal-ion transporter. Nature 388(6641):482–488
57. Ward PP, Paz E, Conneely OM (2005) Multifunctional roles of lactoferrin: a critical overview. Cell Mol Life Sci 62(22):2540–2548
58. Troost FJ, Steijns J, Saris WH, Brummer RJ (2001) Gastric digestion of bovine lactoferrin in vivo in adults. J Nutr 131(8):2101–2104
59. Fransson GB, Lonnerdal B (1980) Iron in human milk. J Pediatr 96(3 Pt 1):380–384
60. Harada E, Araki Y, Furumura E, Takeuchi T et al (2002) Characteristic transfer of colostrum-derived biologically active substances into cerebrospinal fluid via blood in natural suckling neonatal pigs. J Vet Med A Physiol Pathol Clin Med 49(7):358–364
61. Togawa J, Nagase H, Tanaka K, Inamori M et al (2002) Lactoferrin reduces colitis in rats via modulation of the immune system and correction of cytokine imbalance. Am J Physiol Gastrointest Liver Physiol 283(1):G187–G195
62. Wakabayashi H, Kuwata H, Yamauchi K, Teraguchi S et al (2004) No detectable transfer of dietary lactoferrin or its functional fragments to portal blood in healthy adult rats. Biosci Biotechnol Biochem 68(4):853–860
63. Teraguchi S, Wakabayashi H, Kuwata H, Yamauchi K et al (2004) Protection against infections by oral lactoferrin: evaluation in animal models. Biometals 17(3):231–234
64. Bennett RM, Kokocinski T (1979) Lactoferrin turnover in man. Clin Sci (Lond) 57(5):453–460
65. Debanne MT, Regoeczi E, Sweeney GD, Krestynski F (1985) Interaction of human lactoferrin with the rat liver. Am J Physiol 248(4 Pt 1):G463–G469
66. Hutchens TW, Henry JF, Yip TT, Hachey DL et al (1991) Origin of intact lactoferrin and its DNA-binding fragments found in the urine of human milk-fed preterm infants. Evaluation by stable isotopic enrichment. Pediatr Res 29(3):243–250
67. Van Snick JL, Masson PL, Heremans JF (1974) The involvement of lactoferrin in the hypos-ideremia of acute inflammation. J Exp Med 140(4):1068–1084

68. Van Snick JL, Masson PL (1976) The binding of human lactoferrin to mouse peritoneal cells. J Exp Med 144(6):1568–1580

69. Kawamata T, Tooyama I, Yamada T, Walker DG et al (1993) Lactotransferrin immunocytochemistry in Alzheimer and normal human brain. Am J Pathol 142(5):1574–1585

70. Leveugle B, Spik G, Perl DP, Bouras C et al (1994) The iron-binding protein lactotransferrin is present in pathologic lesions in a variety of neurodegenerative disorders: a comparative immunohistochemical analysis. Brain Res 650(1):20–31

71. Fillebeen C, Descamps L, Dehouck MP, Fenart L et al (1999) Receptor-mediated transcytosis of lactoferrin through the blood-brain barrier. J Biol Chem 274(11):7011–7017

72. Ji B, Maeda J, Higuchi M, Inoue K et al (2006) Pharmacokinetics and brain uptake of lactoferrin in rats. Life Sci 78(8):851–855

73. Thorne RG, Lakkaraju A, Rodriguez-Boulan E, Nicholson C (2008) In vivo diffusion of lactoferrin in brain extracellular space is regulated by interactions with heparan sulfate. Proc Natl Acad Sci USA 105(24):8416–8421

74. van Hooijdonk AC, Kussendrager KD, Steijns JM (2000) In vivo antimicrobial and antiviral activity of components in bovine milk and colostrum involved in non-specific defence. Br J Nutr 84(Suppl 1):S127–S134

75. Jenssen H, Hancock RE (2009) Antimicrobial properties of lactoferrin. Biochimie 91(1):19–29

76. Griffiths E, Humphreys J (1977) Bacteriostatic effect of human milk and bovine colostrum on Escherichia coli: importance of bicarbonate. Infect Immun 15(2):396–401

77. Ellison RT 3rd, Giehl TJ, LaForce FM (1988) Damage of the outer membrane of enteric gram-negative bacteria by lactoferrin and transferrin. Infect Immun 56(11):2774–2781

78. Ellison RT 3rd, Giehl TJ (1991) Killing of gram-negative bacteria by lactoferrin and lysozyme. J Clin Invest 88(4):1080–1091

79. Naidu SS, Svensson U, Kishore AR, Naidu AS (1993) Relationship between antibacterial activity and porin binding of lactoferrin in Escherichia coli and Salmonella typhimurium. Antimicrob Agents Chemother 37(2):240–245

80. Erdei J, Forsgren A, Naidu AS (1994) Lactoferrin binds to porins OmpF and OmpC in Escherichia coli. Infect Immun 62(4):1236–1240

81. Leitch EC, Willcox MD (1999) Elucidation of the antistaphylococcal action of lactoferrin and lysozyme. J Med Microbiol 48(9):867–871

82. Rogers HJ, Synge C (1978) Bacteriostatic effect of human milk on Escherichia coli: the role of IgA. Immunology 34(1):19–28

83. Wakabayashi H, Abe S, Okutomi T, Tansho S et al (1996) Cooperative anti-Candida effects of lactoferrin or its peptides in combination with azole antifungal agents. Microbiol Immunol 40(11):821–825

84. Britigan BE, Hayek MB, Doebbeling BN, Fick RB Jr (1993) Transferrin and lactoferrin undergo proteolytic cleavage in the Pseudomonas aeruginosa-infected lungs of patients with cystic fibrosis. Infect Immun 61(12):5049–5055

85. Rogan MP, Taggart CC, Greene CM, Murphy PG et al (2004) Loss of microbicidal activity and increased formation of biofilm due to decreased lactoferrin activity in patients with cystic fibrosis. J Infect Dis 190(7):1245–1253

86. Singh PK, Parsek MR, Greenberg EP, Welsh MJ (2002) A component of innate immunity prevents bacterial biofilm development. Nature 417(6888):552–555

87. Orsi N (2004) The antimicrobial activity of lactoferrin: current status and perspectives. Biometals 17(3):189–196

88. Petschow BW, Talbott RD, Batema RP (1999) Ability of lactoferrin to promote the growth of Bifidobacterium spp. in vitro is independent of receptor binding capacity and iron saturation level. J Med Microbiol 48(6):541–549

89. Mickelsen PA, Blackman E, Sparling PF (1982) Ability of Neisseria gonorrhoeae, Neisseria meningitidis, and commensal Neisseria species to obtain iron from lactoferrin. Infect Immun 35(3):915–920

90. Schryvers AB, Morris LJ (1988) Identification and characterization of the human lactoferrin-binding protein from Neisseria meningitidis. Infect Immun 56(5):1144–1149

 91. Beddek AJ, Schryvers AB (2010) The lactoferrin receptor complex in Gram negative bacteria. Biometals 23(3):377–386
 92. Bellamy W, Takase M, Yamauchi K, Wakabayashi H et al (1992) Identification of the bactericidal domain of lactoferrin. Biochim Biophys Acta 1121(1–2):130–136
 93. Yamauchi K, Tomita M, Giehl TJ, Ellison RT 3rd (1993) Antibacterial activity of lactoferrin and a pepsin-derived lactoferrin peptide fragment. Infect Immun 61(2):719–728
 94. Tomita M, Takase M, Wakabayashi H, Bellamy W (1994) Antimicrobial peptides of lactoferrin. Adv Exp Med Biol 357:209–218
 95. Yoo YC, Watanabe S, Watanabe R, Hata K et al (1997) Bovine lactoferrin and lactoferricin, a peptide derived from bovine lactoferrin, inhibit tumor metastasis in mice. Jpn J Cancer Res 88(2):184–190
 96. Miyauchi H, Hashimoto S, Nakajima M, Shinoda I et al (1998) Bovine lactoferrin stimulates the phagocytic activity of human neutrophils: identification of its active domain. Cell Immunol 187(1):34–37
 97. Shinoda I, Takase M, Fukuwatari Y, Shimamura S et al (1996) Effects of lactoferrin and lactoferricin on the release of interleukin 8 from human polymorphonuclear leukocytes. Biosci Biotechnol Biochem 60(3):521–523
 98. Japelj B, Pristovsek P, Majerle A, Jerala R (2005) Structural origin of endotoxin neutralization and antimicrobial activity of a lactoferrin-based peptide. J Biol Chem 280(17): 16955–16961
 99. Chapple DS, Joannou CL, Mason DJ, Shergill JK et al (1998) A helical region on human lactoferrin. Its role in antibacterial pathogenesis. Adv Exp Med Biol 443:215–220
100. van der Kraan MI, Groenink J, Nazmi K, Veerman EC et al (2004) Lactoferrampin: a novel antimicrobial peptide in the N1-domain of bovine lactoferrin. Peptides 25(2):177–183
101. Shimizu K, Matsuzawa H, Okada K, Tazume S et al (1996) Lactoferrin-mediated protection of the host from murine cytomegalovirus infection by a T-cell-dependent augmentation of natural killer cell activity. Arch Virol 141(10):1875–1889
102. Marchetti M, Longhi C, Conte MP, Pisani S et al (1996) Lactoferrin inhibits herpes simplex virus type 1 adsorption to Vero cells. Antiviral Res 29(2–3):221–231
103. Andersen JH, Osbakk SA, Vorland LH, Traavik T et al (2001) Lactoferrin and cyclic lactoferricin inhibit the entry of human cytomegalovirus into human fibroblasts. Antiviral Res 51(2): 141–149
104. Marr AK, Jenssen H, Moniri MR, Hancock RE et al (2009) Bovine lactoferrin and lactoferricin interfere with intracellular trafficking of Herpes simplex virus-1. Biochimie 91(1):160–164
105. Superti F, Ammendolia MG, Valenti P, Seganti L (1997) Antirotaviral activity of milk proteins: lactoferrin prevents rotavirus infection in the enterocyte-like cell line HT-29. Med Microbiol Immunol 186(2–3):83–91
106. Viani RM, Gutteberg TJ, Lathey JL, Spector SA (1999) Lactoferrin inhibits HIV-1 replication in vitro and exhibits synergy when combined with zidovudine. AIDS 13(10):1273–1274
107. Berkhout B, Floris R, Recio I, Visser S (2004) The antiviral activity of the milk protein lactoferrin against the human immunodeficiency virus type 1. Biometals 17(3):291–294
108. Wu HF, Monroe DM, Church FC (1995) Characterization of the glycosaminoglycan-binding region of lactoferrin. Arch Biochem Biophys 317(1):85–92
109. van der Strate BW, Beljaars L, Molema G, Harmsen MC et al (2001) Antiviral activities of lactoferrin. Antiviral Res 52(3):225–239
110. Puddu P, Borghi P, Gessani S, Valenti P et al (1998) Antiviral effect of bovine lactoferrin saturated with metal ions on early steps of human immunodeficiency virus type 1 infection. Int J Biochem Cell Biol 30(9):1055–1062
111. Yi M, Kaneko S, Yu DY, Murakami S (1997) Hepatitis C virus envelope proteins bind lactoferrin. J Virol 71(8):5997–6002
112. Nozaki A, Ikeda M, Naganuma A, Nakamura T et al (2003) Identification of a lactoferrin-derived peptide possessing binding activity to hepatitis C virus E2 envelope protein. J Biol Chem 278(12):10162–10173

113. Leon-Sicairos N, Lopez-Soto F, Reyes-Lopez M, Godinez-Vargas D et al (2006) Amoebicidal activity of milk, apo-lactoferrin, sIgA and lysozyme. Clin Med Res 4(2):106–113
114. Tanaka T, Omata Y, Narisawa M, Saito A et al (1997) Growth inhibitory effect of bovine lactoferrin on Toxoplasma gondii tachyzoites in murine macrophages: role of radical oxygen and inorganic nitrogen oxide in Toxoplasma growth-inhibitory activity. Vet Parasitol 68(1–2):27–33
115. Cirioni O, Giacometti A, Barchiesi F, Scalise G (2000) Inhibition of growth of Pneumocystis carinii by lactoferrins alone and in combination with pyrimethamine, clarithromycin and minocycline. J Antimicrob Chemother 46(4):577–582
116. Iwamaru Y, Shimizu Y, Imamura M, Murayama Y et al (2008) Lactoferrin induces cell surface retention of prion protein and inhibits prion accumulation. J Neurochem 107(3):636–646
117. Actor JK, Hwang SA, Kruzel ML (2009) Lactoferrin as a natural immune modulator. Curr Pharm Des 15(17):1956–1973
118. Puddu P, Valenti P, Gessani S (2009) Immunomodulatory effects of lactoferrin on antigen presenting cells. Biochimie 91(1):11–18
119. Breton-Gorius J, Mason DY, Buriot D, Vilde JL et al (1980) Lactoferrin deficiency as a consequence of a lack of specific granules in neutrophils from a patient with recurrent infections. Detection by immunoperoxidase staining for lactoferrin and cytochemical electron microscopy. Am J Pathol 99(2):413–428
120. Guillen C, McInnes IB, Vaughan DM, Kommajosyula S et al (2002) Enhanced Th1 response to Staphylococcus aureus infection in human lactoferrin-transgenic mice. J Immunol 168(8):3950–3957
121. Schaible UE, Collins HL, Priem F, Kaufmann SH (2002) Correction of the iron overload defect in beta-2-microglobulin knockout mice by lactoferrin abolishes their increased susceptibility to tuberculosis. J Exp Med 196(11):1507–1513
122. Chodaczek G, Zimecki M, Lukasiewicz J, Lugowski C (2006) A complex of lactoferrin with monophosphoryl lipid A is an efficient adjuvant of the humoral and cellular immune response in mice. Med Microbiol Immunol 195(4):207–216
123. Deriy LV, Chor J, Thomas LL (2000) Surface expression of lactoferrin by resting neutrophils. Biochem Biophys Res Commun 275(1):241–246
124. Gahr M, Speer CP, Damerau B, Sawatzki G (1991) Influence of lactoferrin on the function of human polymorphonuclear leukocytes and monocytes. J Leukoc Biol 49(5):427–433
125. Bournazou I, Pound JD, Duffin R, Bournazos S et al (2009) Apoptotic human cells inhibit migration of granulocytes via release of lactoferrin. J Clin Invest 119(1):20–32
126. de la Rosa G, Yang D, Tewary P, Varadhachary A et al (2008) Lactoferrin acts as an alarmin to promote the recruitment and activation of APCs and antigen-specific immune responses. J Immunol 180(10):6868–6876
127. Birgens HS, Hansen NE, Karle H, Kristensen LO (1983) Receptor binding of lactoferrin by human monocytes. Br J Haematol 54(3):383–391
128. Eda S, Kikugawa K, Beppu M (1996) Binding characteristics of human lactoferrin to the human monocytic leukemia cell line THP-1 differentiated into macrophages. Biol Pharm Bull 19(2):167–175
129. Wakabayashi H, Takakura N, Teraguchi S, Tamura Y (2003) Lactoferrin feeding augments peritoneal macrophage activities in mice intraperitoneally injected with inactivated Candida albicans. Microbiol Immunol 47(1):37–43
130. Sorimachi K, Akimoto K, Hattori Y, Ieiri T et al (1997) Activation of macrophages by lactoferrin: secretion of TNF-alpha, IL-8 and NO. Biochem Mol Biol Int 43(1):79–87
131. Damiens E, Mazurier J, el Yazidi I, Masson M (1998) Effects of human lactoferrin on NK cell cytotoxicity against haematopoietic and epithelial tumour cells. Biochim Biophys Acta 1402(3):277–287
132. Shau H, Kim A, Golub SH (1992) Modulation of natural killer and lymphokine-activated killer cell cytotoxicity by lactoferrin. J Leukoc Biol 51(4):343–349
133. Kuhara T, Yamauchi K, Tamura Y, Okamura H (2006) Oral administration of lactoferrin increases NK cell activity in mice via increased production of IL-18 and type I IFN in the small intestine. J Interferon Cytokine Res 26(7):489–499

134. Wang WP, Iigo M, Sato J, Sekine K et al (2000) Activation of intestinal mucosal immunity in tumor-bearing mice by lactoferrin. Jpn J Cancer Res 91(10):1022–1027
135. Iigo M, Kuhara T, Ushida Y, Sekine K et al (1999) Inhibitory effects of bovine lactoferrin on colon carcinoma 26 lung metastasis in mice. Clin Exp Metastasis 17(1):35–40
136. Sendide K, Deghmane AE, Pechkovsky D, Av-Gay Y et al (2005) Mycobacterium bovis BCG attenuates surface expression of mature class II molecules through IL-10-dependent inhibition of cathepsin S. J Immunol 175(8):5324–5332
137. Fulton SA, Reba SM, Pai RK, Pennini M et al (2004) Inhibition of major histocompatibility complex II expression and antigen processing in murine alveolar macrophages by Mycobacterium bovis BCG and the 19-kilodalton mycobacterial lipoprotein. Infect Immun 72(4):2101–2110
138. Wilk KM, Hwang SA, Actor JK (2007) Lactoferrin modulation of antigen-presenting-cell response to BCG infection. Postepy Hig Med Dosw (Online) 61:277–282
139. Hwang SA, Kruzel ML, Actor JK (2009) Influence of bovine lactoferrin on expression of presentation molecules on BCG-infected bone marrow derived macrophages. Biochimie 91(1):76–85
140. Curran CS, Demick KP, Mansfield JM (2006) Lactoferrin activates macrophages via TLR4-dependent and -independent signaling pathways. Cell Immunol 242(1):23–30
141. Mincheva-Nilsson L, Hammarstrom S, Hammarstrom ML (1997) Activated human gamma delta T lymphocytes express functional lactoferrin receptors. Scand J Immunol 46(6):609–618
142. Zimecki M, Mazurier J, Machnicki M, Wieczorek Z et al (1991) Immunostimulatory activity of lactotransferrin and maturation of CD4- CD8- murine thymocytes. Immunol Lett 30(1):119–123
143. Dhennin-Duthille I, Masson M, Damiens E, Fillebeen C et al (2000) Lactoferrin upregulates the expression of CD4 antigen through the stimulation of the mitogen-activated protein kinase in the human lymphoblastic T Jurkat cell line. J Cell Biochem 79(4):583–593
144. Sfeir RM, Dubarry M, Boyaka PN, Rautureau M et al (2004) The mode of oral bovine lactoferrin administration influences mucosal and systemic immune responses in mice. J Nutr 134(2):403–409
145. Wakabayashi H, Takakura N, Yamauchi K, Tamura Y (2006) Modulation of immunity-related gene expression in small intestines of mice by oral administration of lactoferrin. Clin Vaccine Immunol 13(2):239–245
146. Artym J, Zimecki M, Kruzel ML (2003) Reconstitution of the cellular immune response by lactoferrin in cyclophosphamide-treated mice is correlated with renewal of T cell compartment. Immunobiology 207(3):197–205
147. Actor JK, Hwang SA, Olsen M, Zimecki M et al (2002) Lactoferrin immunomodulation of DTH response in mice. Int Immunopharmacol 2(4):475–486
148. Fischer R, Debbabi H, Dubarry M, Boyaka P et al (2006) Regulation of physiological and pathological Th1 and Th2 responses by lactoferrin. Biochem Cell Biol 84(3):303–311
149. Wakabayashi H, Kurokawa M, Shin K, Teraguchi S et al (2004) Oral lactoferrin prevents body weight loss and increases cytokine responses during herpes simplex virus type 1 infection of mice. Biosci Biotechnol Biochem 68(3):537–544
150. Ishii K, Takamura N, Shinohara M, Wakui N et al (2003) Long-term follow-up of chronic hepatitis C patients treated with oral lactoferrin for 12 months. Hepatol Res 25(3):226–233
151. Kuhara T, Iigo M, Itoh T, Ushida Y et al (2000) Orally administered lactoferrin exerts an antimetastatic effect and enhances production of IL-18 in the intestinal epithelium. Nutr Cancer 38(2):192–199
152. Okamoto I, Kohno K, Tanimoto T, Ikegami H et al (1999) Development of CD8+ effector T cells is differentially regulated by IL-18 and IL-12. J Immunol 162(6):3202–3211
153. Spadaro M, Caorsi C, Ceruti P, Varadhachary A et al (2008) Lactoferrin, a major defense protein of innate immunity, is a novel maturation factor for human dendritic cells. FASEB J 22(8):2747–2757

154. Hwang SA, Actor JK (2009) Lactoferrin modulation of BCG-infected dendritic cell functions. Int Immunol 21(10):1185–1197

155. Bennett RM, Davis J (1981) Lactoferrin binding to human peripheral blood cells: an interaction with a B-enriched population of lymphocytes and a subpopulation of adherent mononuclear cells. J Immunol 127(3):1211–1216

156. Zimecki M, Mazurier J, Spik G, Kapp JA (1995) Human lactoferrin induces phenotypic and functional changes in murine splenic B cells. Immunology 86(1):122–127

157. Debbabi H, Dubarry M, Rautureau M, Tome D (1998) Bovine lactoferrin induces both mucosal and systemic immune response in mice. J Dairy Res 65(2):283–293

158. Artym J, Zimecki M, Kruzel ML (2004) Effect of lactoferrin on the methotrexate-induced suppression of the cellular and humoral immune response in mice. Anticancer Res 24(6):3831–3836

159. Choi BK, Actor JK, Rios S, d'Anjou M et al (2008) Recombinant human lactoferrin expressed in glycoengineered Pichia pastoris: effect of terminal N-acetylneuraminic acid on in vitro secondary humoral immune response. Glycoconj J 25(6):581–593

160. Kruzel ML, Harari Y, Chen CY, Castro GA (2000) Lactoferrin protects gut mucosal integrity during endotoxemia induced by lipopolysaccharide in mice. Inflammation 24(1):33–44

161. Zhang GH, Mann DM, Tsai CM (1999) Neutralization of endotoxin in vitro and in vivo by a human lactoferrin-derived peptide. Infect Immun 67(3):1353–1358

162. Sugi K, Saitoh O, Hirata I, Katsu K (1996) Fecal lactoferrin as a marker for disease activity in inflammatory bowel disease: comparison with other neutrophil-derived proteins. Am J Gastroenterol 91(5):927–934

163. Pereira SP, Rhodes JM, Campbell BJ, Kumar D et al (1998) Biliary lactoferrin concentrations are increased in active inflammatory bowel disease: a factor in the pathogenesis of primary sclerosing cholangitis? Clin Sci (Lond) 95(5):637–644

164. Uchida K, Matsuse R, Tomita S, Sugi K et al (1994) Immunochemical detection of human lactoferrin in feces as a new marker for inflammatory gastrointestinal disorders and colon cancer. Clin Biochem 27(4):259–264

165. Zweiman B, Kucich U, Shalit M, Von Allmen C et al (1990) Release of lactoferrin and elastase in human allergic skin reactions. J Immunol 144(10):3953–3960

166. Bennett RM, Eddie-Quartey AC, Holt PJ (1973) Lactoferrin–an iron binding protein in synovial fluid. Arthritis Rheum 16(2):186–190

167. Haversen L, Ohlsson BG, Hahn-Zoric M, Hanson LA et al (2002) Lactoferrin down-regulates the LPS-induced cytokine production in monocytic cells via NF-kappa B. Cell Immunol 220(2):83–95

168. Slater K, Fletcher J (1987) Lactoferrin derived from neutrophils inhibits the mixed lympho-cyte reaction. Blood 69(5):1328–1333

169. Machnicki M, Zimecki M, Zagulski T (1993) Lactoferrin regulates the release of tumour necrosis factor alpha and interleukin 6 in vivo. Int J Exp Pathol 74(5):433–439

170. Crouch SP, Slater KJ, Fletcher J (1992) Regulation of cytokine release from mononuclear cells by the iron-binding protein lactoferrin. Blood 80(1):235–240

171. Mattsby-Baltzer I, Roseanu A, Motas C, Elverfors J et al (1996) Lactoferrin or a fragment thereof inhibits the endotoxin-induced interleukin-6 response in human monocytic cells. Pediatr Res 40(2):257–262

172. Sugiyama A, Sato A, Shimizu H, Ando K et al (2010) PEGylated lactoferrin enhances its hepatoprotective effects on acute liver injury induced by D-galactosamine and lipopolysac-charide in rats. J Vet Med Sci 72(2):173–180

173. Baveye S, Elass E, Fernig DG, Blanquart C et al (2000) Human lactoferrin interacts with soluble CD14 and inhibits expression of endothelial adhesion molecules, E-selectin and ICAM-1, induced by the CD14-lipopolysaccharide complex. Infect Immun 68(12):6519–6525

174. Elass E, Masson M, Mazurier J, Legrand D (2002) Lactoferrin inhibits the lipopolysaccharide-induced expression and proteoglycan-binding ability of interleukin-8 in human endothelial cells. Infect Immun 70(4):1860–1866

175. Wright SD, Ramos RA, Tobias PS, Ulevitch RJ et al (1990) CD14, a receptor for complexes of lipopolysaccharide (LPS) and LPS binding protein. Science 249(4975):1431–1433
176. Van Amersfoort ES, Van Berkel TJ, Kuiper J (2003) Receptors, mediators, and mechanisms involved in bacterial sepsis and septic shock. Clin Microbiol Rev 16(3):379–414
177. Sanchez L, Calvo M, Brock JH (1992) Biological role of lactoferrin. Arch Dis Child 67(5):657–661
178. McCombs JL, Teng CT, Pentecost BT, Magnuson VL et al (1988) Chromosomal localization of human lactotransferrin gene (LTF) by in situ hybridization. Cytogenet Cell Genet 47(1–2):16–17
179. Kim SJ, Yu DY, Pak KW, Jeong S et al (1998) Structure of the human lactoferrin gene and its chromosomal localization. Mol Cells 8(6):663–668
180. Teng CT, Pentecost BT, Marshall A, Solomon A et al (1987) Assignment of the lactotransferrin gene to human chromosome 3 and to mouse chromosome 9. Somat Cell Mol Genet 13(6):689–693
181. Kang JF, Li XL, Zhou RY, Li LH et al (2008) Bioinformatics analysis of lactoferrin gene for several species. Biochem Genet 46(5–6):312–322
182. Pentecost BT, Teng CT (1987) Lactotransferrin is the major estrogen inducible protein of mouse uterine secretions. J Biol Chem 262(21):10134–10139
183. Liu Y, Yang N, Teng CT (1993) COUP-TF acts as a competitive repressor for estrogen receptor-mediated activation of the mouse lactoferrin gene. Mol Cell Biol 13(3):1836–1846
184. Teng CT (2006) Factors regulating lactoferrin gene expression. Biochem Cell Biol 84(3):263–267
185. Geng K, Li Y, Bezault J, Furmanski P (1998) Induction of lactoferrin expression in murine ES cells by retinoic acid and estrogen. Exp Cell Res 245(1):214–220
186. Lee MO, Liu Y, Zhang XK (1995) A retinoic acid response element that overlaps an estrogen response element mediates multihormonal sensitivity in transcriptional activation of the lactoferrin gene. Mol Cell Biol 15(8):4194–4207
187. Hasmall S, Orphanides G, James N, Pennie W et al (2002) Downregulation of lactoferrin by PPARalpha ligands: role in perturbation of hepatocyte proliferation and apoptosis. Toxicol Sci 68(2):304–313
188. Verbeek W, Lekstrom-Himes J, Park DJ, Dang PM et al (1999) Myeloid transcription factor C/EBPepsilon is involved in the positive regulation of lactoferrin gene expression in neutrophils. Blood 94(9):3141–3150
189. Yamanaka R, Barlow C, Lekstrom-Himes J, Castilla LH et al (1997) Impaired granulopoiesis, myelodysplasia, and early lethality in CCAAT/enhancer binding protein epsilon-deficient mice. Proc Natl Acad Sci USA 94(24):13187–13192
190. Green MR, Pastewka JV (1978) Lactoferrin is a marker for prolactin response in mouse mammary explants. Endocrinology 103(4):151–203
191. Nelson KG, Takahashi T, Bossert NL, Walmer DK et al (1991) Epidermal growth factor replaces estrogen in the stimulation of female genital-tract growth and differentiation. Proc Natl Acad Sci USA 88(1):21–25
192. Ward PP, Mendoza-Meneses M, Mulac-Jericevic B, Cunningham GA et al (1999) Restricted spatiotemporal expression of lactoferrin during murine embryonic development. Endocrinology 140(4):1852–1860
193. Breton M, Mariller C, Benaissa M, Caillaux K et al (2004) Expression of delta-lactoferrin induces cell cycle arrest. Biometals 17(3):325–329
194. Mariller C, Benaissa M, Hardiville S, Breton M et al (2007) Human delta-lactoferrin is a transcription factor that enhances Skp1 (S-phase kinase-associated protein) gene expression. FEBS J 274(8):2038–2053
195. Kanyshkova TG, Babina SE, Semenov DV, Isaeva N et al (2003) Multiple enzymic activities of human milk lactoferrin. Eur J Biochem 270(16):3353–3361

Chapter 4
Lactoferrin as a Signaling Mediator

Abstract Lactoferrin is a metal-binding protein, secreted from glandular epithelial cells and neutrophils. As well as other growth factors and cytokines, it plays a role for regulation of cell behavior by interacting with target cells and molecules. At the surface of the cells, the sulfated chain of proteoglycans is considered as primary lactoferrin binding site. The initial binding to proteoglycans can induce lactoferrin interaction to specific receptors, such as intelectin, LDL-receptor related protein (LRP), nucleolin and CD14. Lymphocyte expresses lactoferrin receptor which molecular weight is about 105 kDa. However, molecular nature of the lymphocye lactoferrin receptor is unknown. Some of lactoferrin receptors are involved in receptor-mediated uptake of lactoferrin. Lactoferrin acts as an anabolic factor for skeletal tissue and promotes the growth and differentiation of osteoblasts and chondrocytes. Lactoferrin antagonizes bone resorption by inhibiting osteoclastic differentiation. It inhibits the tumor cell growth by regulating the expression of and phosphorylation of cyclin-dependent kinase inhibitors (CKIs). The signal transduction pathways induced by lactoferrin is partially understood.

Keywords Cell cycle • Intelectin • LDL-receptor related protein (LRP) • Nucleolin • Osteoblasts

4.1 Introduction

Lactoferrin, a metal-binding glycoprotein secreted from glandular epithelial cells and neutrophils, can regulate the proliferation and differentiation of many types of cells. The effects of lactoferrin are dependent on the cell type and state of differentiation. Lactoferrin exerts its physiological functions by binding to specific lactoferrin receptors on target cells. Lactoferrin is known as highly basic protein, and interacts with target cells or molecules by means of positive charge on their surface [1]. At the surface of the cells, the sulfated chain of proteoglycans is regarded as primary lactoferrin binding site. The initial binding to proteoglycans can induce lactoferrin interaction to specific receptors [2]. Different types of tissues and cells express different lactoferrin receptors [3]. The diverse functions can result from activation of different receptors and signal transduction pathways, and also from differences in the expression levels of receptors. Intestinal lactoferrin receptor was identified as intelectin. The Low-density lipoprotein receptor related protein-1 (LRP-1) acts as a lactoferrin signaling receptor in fibroblasts and osteoblasts. Hepatocyte expresses two types of lactoferrin receptor, LRP-1 and asialoglycoprotein receptor (ASGP receptor). RNA binding protein nucleolin is characterized as one of the lymphocyte lactoferrin receptor. CD14 has been postulated as monocyte lactoferrin receptor. In this article, recent advances in the understanding the function of lactoferrin at cellular and molecular level are summarized.

4.2 Intestinal Lactoferrin Receptor (Intelectin)

As lactoferrin was first identified as host defense component of milk and colostrum, lactoferrin receptor was first discovered in small intestine. The first intestinal lactoferrin receptor has been purified from mouse [4]. The purified receptor has a molecular weight of ~130 kDa. Deglycosylation of the lactoferrin receptor results in a decrease of the molecular weight to approximately ~105 kDa. cDNA encoding the intestinal lactoferrin receptor was identified as mouse intelectin [5]. It encodes 313 amino acids, and deduced molecular weight is only 34 kDa. However, the recombinant mouse intelectin expressed in baculovirus expression system has a molecular weight of 102 kDa under reducing conditions in SDS-PAGE [5]. The discrepancy of molecular size suggests that mouse intelectin is a trimer of the 34-kDa proteins.

Human intestinal lactoferrin receptor has an apparent molecular weight of 114 kDa [6]. The molecular weight of deglycosylated single subunit is 34 kDa, suggesting that human intestinal lactoferrin receptor exists as trimer in non-reducing conditions. Amino acid sequence deduced from cDNA showed 81% identity with mouse intelectin. Thereby, human intestinal lactoferrin receptor was identified as human intelectin. Intelectin is a novel soluble lectin that recognizes galactofuranose in carbohydrate chains of bacterial cells [7]. Study using recombinant human intelectin expressed in baculovirus expression system indicates that calcium is required for the interaction between human lactoferrin and recombinant human intelectin. In human, an abundant intelectin mRNA expression is observed in salivary gland, heart, skeletal muscle, testis, adrenal gland and pancreas. In mouse, intelectin mRNA is found in small intestine, spleen and lung. In fetal tissue, intelectin is abundantly expressed in small intestine [8]. The biological role of the intelectin on lactoferrin metabolism is not fully understood.

4.3 LDL Receptor Related Protein (LRP)

Lactoferrin can interact with LRP-1 and LRP-2 (LDL receptor related protein 1 and 2) [9]. LRP-1 (CD91, $\alpha2$ macroglobulin receptor) is a membrane glycoprotein that is a member of the low-density lipoprotein (LDL) receptor family [10, 11]. LRP-1 is abundantly expressed on hepatocytes, neurons, smooth muscle cells, osteoblasts and fibroblasts and consists of an extracellular 515-kDa heavy chain and 85-kDa light chain that spans the membrane (Fig. 4.1) [10]. The extracellular domain of LRP contains four ligand-binding clusters denoted I to IV. The light chain contains a single transmembrane domain and two NPXY motifs, which function in recruitment of adaptor proteins and clathrin for endocytosis. The heavy chain is noncovalently associated with the light chain on the cell surface. LRP-2 (Megalin/gp300) expression is mainly found in absorptive epithelia, such as renal proximal tubules, gallbladder or mammary epithelia [12]. Similar to other members in the LDL receptor family, LRP-1 is an endocytotic receptor and participates in the uptake of lipoproteins containing triglyceride and cholesterol by hepatocytes. However, the broad range of its ligand diversity and lethality of LRP-1 conventional knockout mice suggests that LRP-1 is involved in diverse physiological and pathological processes other than just lipoprotein metabolism [13]. These processes may include cell

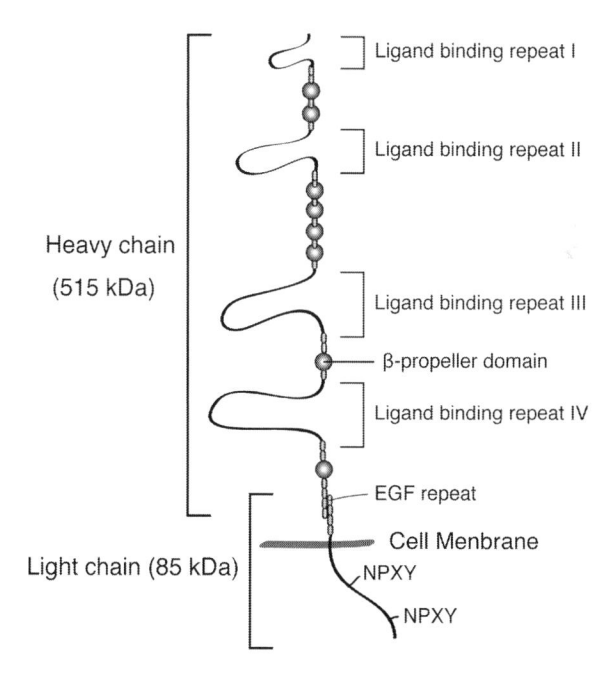

Fig. 4.1 Structure of low-density lipoprotein receptor-related protein 1 (LRP-1). LRP-1 consists of 515 kDa heavy chain and 85 kDa light chain. The heavy chain contains four ligand-binding repeats. The light chain contains a single transmembrane domain and two NPXY motifs, which function in recruitment of adaptor proteins and clathrin for endocytosis. The heavy chain is noncovalently associated with the light chain on the cell surface

migration, fibrinolysis, thrombosis, and atherosclerosis. It has been reported that the binding of various ligands (e.g. urokinase plasminogen activator (uPA), apolipopro-teinE (apoE), and β-defensin) to LRP-1 regulates the migration in cultured cells and the contraction of smooth muscle cells. Lactoferrin can interact with the second and fourth cluster of complement type repeats of lactoferrin [14]. Translocation of lac-toferrin across blood brain barrier is mediated by LRP-1 [15]. In fibroblasts and osteoblasts, LRP-1 acts as signaling receptor for lactoferrin by converting lactoferrin-binding to activation of ERK1/2 [16, 17]. The endocytic activity of LRP-1 is not required for its role as signaling receptor. Similarly, LRP-1 is involved in lactoferrin-enhanced motility of keratinocytes [18].

4.4   Hepatocyte Lactoferrin Receptor

Intravenous injected lactoferrin is rapidly removed from circulation mainly due to receptor-mediated endocytosis of hepatocytes [19, 20]. Lactoferrin inhibits the LRP-mediated uptake of apolipoprotein E (Apo E) by liver, suggesting that LRP-1 is responsible for the hepatocyte-mediated lactoferrin uptake [21]. Removal of 14 N terminal cationic amino acids from human lactoferrin by amino peptidase enhances its affinity for parenchymal liver cells [21]. The N-terminal deleted human lactoferrin can inhibit VLDL uptake by hepatocytes, suggesting that this N-terminal region is not essential for lactoferrin binding to hepatocytes.

In addition to LRP-1, Ca^{2+}-dependent high affinity lactoferrin receptor was iden-tified in rat liver [22]. Lactoferrin can bind the RHL-1 subunit of asialoglycoprotein (ASGP) receptor. ASGP receptor has lectin activity. However, deglycosilated bovine lactoferrin can compete with intact bovine lactoferrin for ASCP binding [23]. Rat hepatocytes are known to internalize lactoferrin by clathrin-mediated endocytosis.

4.5   Nucleolin

Nucleolin is a 105-kDa nucleus RNA-binding protein that is ubiquitously expressed in growing eukaryotic cells. Nucleolin participate in maintenance of chromatin structure, rRNA maturation and ribosome assembly [24]. Nucleolin can shuttle from plasma membranes to the nucleus [25]. Furthermore, nucleolin could function as a cell surface receptor for several ligands, such as matrix laminin-1, midkine, attachment factor J, Apo-B and Apo-E [26–29]. Nucleolin was identified as a major cell surface lactoferrin-binding site in Chinese hamster ovary (CHO) cells. Lactoferrin competes with nucleolin binding for HB-19, a pseudo-peptide that binds the C-terminal end of nucleolin. In MDA-MB-231 human breast cancer cells, nucle-olin co-localizes with lactoferrin on the cell surface and early endosomes, suggest-ing that nucleolin participates in lactoferrin endocytosis [30]. Lactoferrin binding protein of 105 kDa was identified in T-cells and platelets [31–33]. However, it is unlikely that these receptors would be nucleolin, since human lactoferrin residues 28–34 is reported to be essential for its binding to 105 kDa receptor, whereas both N-lobe and C-lobe of lactoferrin can bind nucleolin, and the cationic N-terminal region is dispensable. Heparin sulfate proteoglycans are required for nucleolin-mediated lactoferrin endocytosis [30].

4.6   Monocytes and Macrophage Lactoferrin Receptor

The effects of lactoferrin on regulation of innate immune response are important for the first line host defense against various pathogens. Monocyte is the first mammalian cell which lactoferrin binding activity is identified [34]. Lactoferrin promotes the differentiation of THP-1 cells, human monocytic leukemia cell line, into macrophages, and the binding capacity of lactoferrin is increased along with the differentiation process. Both human and bovine lactoferrin can interact with THP-1 cells. Heparin inhibits specific binding of both human and bovine lactoferrin to THP-1 cells in a dose-dependent manner [35]. Pretreatment of THP-1 cells with $NaClO_3$ to prevent sulphation of surface glycosaminoglycans attenuates lactoferrin binding, and de-N-sulphated heparin did not inhibit binding of lactoferrin to THP-1 cells, suggesting that heparin binding and monocyte/macrophage binding by lactoferrin both involve interactions between basic regions of lactoferrin and sulphate groups [35]. At least four lactoferrin-binding proteins were found in the membrane of THP-1 cells [36].

In monocytes and macrophages, lipopolysaccharide (LPS) induces the productions of pro-inflammatory cytokines by toll-like receptor 4 (TLR4)-dependent mechanism. Monocytes and macrophages express mCD14 (membrane bound CD14) as a LPS receptor. LPS-mCD14 complex on cell surface is recognized by TLR4. Lactoferrin antagonizes the LPS-induced production of the pro-inflammatory cytokines, such as TNFα, IL-1β, IL-6 and IL-8 [37, 38]. Lactoferrin competes with LPS for mCD14 binding [39], suggesting that lactoferrin exerts their biological effect on monocytes by neutralizing LPS effect. One of the lactoferrin receptor in monocyte is likely to be mCD14. In this process, lactoferrin enters the THP-1 cells and inhibits the LPS-induced binding of NF-κB to the TNF-α promoter.

Besides the inhibitory effect of lactoferrin on LPS-induced signaling pathway, lactoferrin can solely induce CD40 expression, and IL-6 production in THP-1 cells [40]. Lactoferrin treatment induces NF-κB, p38 MAPK, ERK and JNK in macrophages. The promoting effect of lactoferrin on CD40 expression is dependent on TLR4 expression, whereas lactoferrin does not require TLR4 expression with respect to IL-6 production [41]. Furthermore, lactoferrin can activate NF-κB, p38 MAPK, ERK and JNK in TLR4-deficient macrophages whereas LPS-induced cell signaling is blocked in the TLR4-deficient cells [41]. These lines of observations suggest that lactoferrin can activate intracellular signaling pathways other than CD14/TLR4 pathway. Indeed, LRP and nucleolin are abundantly expressed in macrophages and are considered as macrophage lactoferrin receptors [42, 43].

4.7 Dendritic Cell Lactoferrin Receptor

While lactoferrin can promote dendritic cell functions as antigen presenting cells, the nature of lactoferrin receptor expressed in dendritic cell is still unknown. However, the effect of lactoferrin on skin Langerhans cell is inhibited by blocking the mannose receptor, suggesting that the mannose receptor is putative lactoferrin binding site for dendritic cells [44]. Moreover, lactoferrin binds to dendritic cell specific ICAM3-grabbing non-integrin (DC-SIGN), a C-type lectin receptor [45]. DC-SIGN-mediated transfer of human immuno-deficiency virus-1 (HIV-1) to CD4+ T-cells is inhibited by bovine lactoferrin [46].

4.8 Lymphocytes Lactoferrin Receptor

Lactoferrin binding to lymphocytes is observed only when they are activated by mitogen [32]. They are found to express lactoferrin receptor which molecular weight is ~105 kDa. The iron saturation of lactoferrin does not affect the lactoferrin bindings to the lymphocyte receptor. However, the effect of lactoferrin on the lymphocyte proliferation is dependent on the iron statues of lactoferrin. According to the study using lactoferrin tryptic fragment, the lactoferrin fragment corresponding to N1 domain of human lactoferrin binds to the lymphocyte lactoferrin receptor with same affinity as intact human lactoferrin. The N2 domain fragment and C-lobe fragment of human lactoferrin does not binds lactoferrin receptor, suggesting that the N1 domain is responsible for lactoferrin-binding to the lymphocyte lactoferrin receptor.

Lactoferrin can bind and incorporate to B-lymphocytes [47]. Lactoferrin promotes the maturation of B-lymphocytes by an increase of surface IgD and complement receptor expression [48].

Lactoferrin receptor expression can be detected in CD4+, CD8+, and γδ T-cells [49]. In human lymphoblastic Jurkat T-cell line, lactoferrin is internalized from cell membrane by endocytic recycling [50]. Lactoferrin increased CD4 expression through the activation of MAPK via 105 kDa lactoferrin receptor [51]. This activation is inhibited by PP1, tyrosine kinase inhibitor, suggesting that non-receptor type tyrosine kinase, such as p56lck, is involved in lactoferrin-induced MAPK activation in T cells. Lactoferrin increases expression of T-cell receptor ζ chain that is a component of the CD3 T-cell receptor complex necessary for T-cell activation and subsequent proliferation [52].

4.9 Cell Cycle Control and Apoptosis Induction

Lactoferrin is involved in the cell cycle control system and apoptosis in a variety of cancer cell lines. Lactoferrin induces cell cycle arrest by regulating the expression or phosphorylation of cell cycle regulators, including cyclins, cyclin-dependent

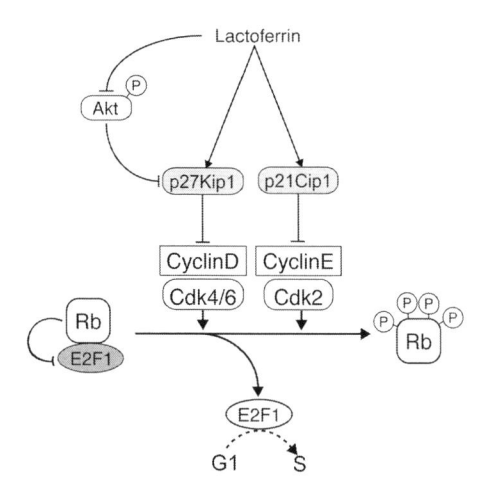

Fig. 4.2 Lactoferrin inhibits cell cycle progression in cancer cells. The CyclinD-Cdk4/6 (cyclin-dependent kinase 4/6) induces G1/S cell cycle progression by phosphorylating Rb (Retinoblastoma protein). Rb phosphorylation causes E2F1 release from Rb-E2F1 complex, and activates S-phase gene expression. The cyclin E/Cdk2 complex regulates entry of cells into S-phase by phosphorylating Rb. Lactoferrin increases the expression of the two Cdk inhibitors (p21Cip1 and p27Kip1) to block Rb phosphorylation. p27Kip1 is inactivated by Akt-mediated phosphorylation. Lactoferrin-induced loss of Akt activity induces the translocation of p27Kip1 into nucleus where it associates with the cyclin-Cdk complex

kinases (CDKs) and cyclin-dependent kinase inhibitors (CKIs), such as p21Cip1 and p27Kip1 (Fig. 4.2). In human breast carcinoma cells (MDA-MB-231), human lactoferrin induces cell cycle arrest at the G_1 to S transition. This lactoferrin-induced arrest is associated with elevation of p21Cip1 expression [53]. The expression of p21Cip1 is increased in cancer cells treated with lactoferrin, or cells transiently transfected lactoferrin gene [54]. Moreover, the lactoferrin-enhanced elevation of p21Cip expression is correlated with reduction in the protein levels of Cdk2 and cyclin E, and in the kinase activities of Cdk2 and CGK4 [53]. In HeLa cells, over-expression of lactoferrin increases protein levels of p21Cip1 by activating NF-κB cascade [55]. Lactoferrin also increases p27Kip1 expression and inhibits growth of cancer cell lines, accompanied by suppression of cyclin E expression [57]. Lactoferrin treatment enhances the expression of p21Cip1 and p27Kip1 in human nasopharyngeal carcinoma cells and induces their cycle arrest [56]. These observations are consisted with previous observations showing that p21Cip1 and p27Kip1 mediates cell cycle arrest in response to a variety of stress stimuli.

The serine/threonine kinase Akt (protein kinase B) participates in cell cycle progression and inhibits apoptosis induced by a variety of stimuli. Akt phosphorylates p27Kip1, leading to inactivation or destruction of p27Kip1. Down regulation of Akt phosphorylation (activation) can be observed in the lactoferrin treated cancer cells [57]. In SGC-7901 human stomach cancer cells, lactoferrin induces apoptosis by suppressing the Akt expression level and Akt phosphorylation (Thr308/Ser473) [58].

The effect of lactoferrin on tumor cell cycle progression and apoptosis can be explained by Rb (Retinoblastoma protein)/E2F1-dependent mechanism [54]. Rb is ubiquitously expressed tumor suppressor gene product and its expression is sustained at significant levels throughout the cell cycle. Rb acts as a repressor of the E2F family of transcription factors. In G_1 phase, Rb is phosphorylated on several serine and threonine residues by distinct cyclin-cdk complexes. The phosphorylation of Rb dissociates E2F1 from Rb, allowing E2F1 to activate S phase gene expression. Lactoferrin treatment increases Rb expression levels in some cancer cell lines including MCF7, HEK293, and Jurkat T cells [54]. Reduction of Rb phosphorylation is observed in lactoferrin-treated cancer cells [56]. Transfection of lactoferrin gene in H1299 cells (human lung carcinoma cells) induces hypophosphorylation of Rb protein.

Lactoferrin-activated apoptotic machinery uses Bcl-2 family proteins to induce cell death. Bcl-2 family consists of pre-apoptotic and anti-apoptotic members. Bad is pro-apoptotic protein belongs to Bcl-2-family. Bad can interact with Bcl-2 in the outer mitochondrial membranes. Bad promotes cell death by nullifying anti-apoptotic activities of Bcl-2. Activation of c-Jun N-terminal kinases (JNK) pathway induces Bad phosphorylation and promotes their mitochondrial translocation to induce apoptosis. In Jurkat leukemia T cell line, lactoferrin induces apoptosis by JNK-dependent mechanism [59]. Lactoferrin promotes the Bcl-2 phosphorylation and interaction of Bax with Bcl-2 [60]. Bovine lactoferricin induces caspase-independent apoptosis in human B-lymphoma cells [61].

On the other hand, relatively low levels of iron-saturated lactoferrin induces Akt phosphorylation (Ser473) and p21Cip1 and p27Kip1 phosphorylation in PI-3K dependent manner, and stimulate S-phase cell cycle entry [62]. Delta-lactoferrin mRNA is the product of alternative splicing of the lactoferrin gene. Delta-lactoferrin expression is found in normal tissues and is reported to be absent from their malignant counterparts. Its expression provokes anti-proliferative effects and induces cell cycle arrest in S-phase [63].

4.10 Angiogenesis

Angiogenesis inhibitors show considerable potential in the treatment of cancer because it is necessary for tumor growth for oxygen and nutrient supply. Human and bovine apo-lactoferrin exerts opposite effects on angiogenesis. Subcutaneously infused apo-human lactoferrin also significantly stimulates vascular endothelial growth factor (VEGF)-mediated angiogenesis. [64]. The promoting effect of human apo-lactoferrin is also observed in chick chorioallantoic membrane assay [65]. Human apo-lactoferrin augments VEGFR2 (KDR/Flk-1) expression in human umbilical vein endothelial cells (HUVECs) that promotes the VEGF-induced proliferation and migration of the endothelial cells [65]. However, holo-human lactoferrin did not affect VEGF-A-mediated angiogenesis [64, 65].

On the other hand, accumulating evidences indicate that bovine lactoferrin has inhibitory effect on angiogenesis *in vitro* and *in vivo*. Orally administered bovine lactoferrin inhibits VEGF(165)-mediated angiogenesis in the rat [66] or angiogenesis in chick embryo chorioallantoic membranes (CAMs) [67]. Bovine lactoferricin inhibits both fibroblast growth factor 2 (FGF-2)- and VEGF(165)-induced angiogenesis in Matrigel plugs implanted in C57BL/6 mice [68]. In mice, both orally and intraperitoneally administered bovine lactoferrin dose-dependently suppresses Lewis lung carcinoma cell-induced angiogenesis in a dorsal air sac assay [67]. FGF-2- or VEGF-induced proliferation of mouse endothelial KOP2.16 cells is suppressed by bovine lactoferrin [67]. In addition, bovine lactoferricin antagonizes the *in vitro* proliferation and migration of HUVECs in response to FGF-2 or VEGF(165) [68]. The mechanism underlying the opposing effects of bovine and human lactoferrin on angiogenesis is unclear.

4.11 Fibroblasts

Lactoferrin induces ERK1/2 and Rho kinase (ROK/ROCK) activation in fibroblasts [69, 70]. LRP-1 expression is found in human dermal fibroblasts [71]. As described above, the mitogenic effect of lactoferrin on fibroblast is dependent on LRP-1 expression [17, 70]. LRP-1 is also involved in lactoferrin-enhanced fibroblast collagen contraction [16].

As well as LRP-1, toll-like receptor 4 (TLR4) can act as lactoferrin-induced signaling receptor in fibroblasts [40]. Lactoferrin induces NF-κB activation in mouse embryonic fibroblasts (MEFs). This activation is impaired in MEFs lacking TLR4 or tumor necrosis factor receptor-associated factor 6 (TRAF 6), a crucial signaling transducer in LPS/TLR signal transduction. However, TRAF 2 and TRAF 5 deletion does not affect lactoferrin-induced NF-κB activation. Therefore, the TLR4-TRAF6 signaling pathway is critical for lactoferrin-induced NF-κB activation.

4.12 Osteogenic Differentiation

Bone homeostasis is maintained by a balance between bone resorption by osteoclasts and bone formation by osteoblasts. Lactoferrin is identified anabolic factor for osteoblasts. Lactoferrin prevents apotosis of osteoblasts induced by serum deprivation [72]. Lactoferrin promotes the proliferation of primary osteoblasts or osteoblast cell line [72]. The expression of osteoblastic phenotypes such as the alkaline phosphatase (ALP) activity, osteocalcin production and calcium deposition are enhanced by lactoferrin treatment during the differentiation of MG63 human osteosarcoma cells [73, 74].

Both LRP-1 and LRP-2 are expressed in obteoblasts. RAP (receptor associated protein) was identified as a high affinity ligand for LRP-1, LRP-2 and VLDL receptor. RAP inhibits ligand binding to the receptors *in vitro* and *in vivo* [75, 76]. The lactoferrin-enhanced ERK1/2 phosphorylation in osteoblast is antagonized by RAP [17]. This observation indicates that both LRP-1 and LRP-2 can act as a signaling receptor that converts lactoferrin-binding to ERK1/2 activation. However, lactoferrin induced activation of phosphatidylinositol-3 kinase (PI-3 K) is not blocked by RAP [77], suggesting that receptors other than LRP-1 act as signaling receptor in osteoblasts and activate PI-3 K signaling pathway. LRP is known as an endocytotic receptor and participates in the uptake of lipoproteins, and plasminogen activator. However, lactoferrin-induced ERK1/2 activation is not blocked by inhibition of lactoferrin endocytosis by hyperosmotic medium, suggesting that endocytic activity of LRP is not required for its role as signaling receptor. In murine osteoblast-like MC3T3-cells, lactoferrin induces VEGF and FGF-2 synthesis by ERK1/2-dependent manner [78]. The detailed mechanisms that involved in lactoferrin-enhanced osteoblastic differentiation are still unknown.

4.13 Osteoclasts

Osteoblasts play a central role in bone formation by synthesizing multiple bone matrix proteins. Moreover, they regulate osteoclast maturation by soluble factors and cognate interaction, resulting in bone resorption. Bovine lactoferrin inhibits the bone resorbing activity of osteoclasts in a rabbit mixed-bone cell culture [79]. Lactoferrin antagonizes osteoclast formation in murine bone marrow culture [72], suggesting that lactoferrin potently inhibits osteoclastogenesis. Whereas lactoferrin shows no effect on bone resorption by isolated mature osteoclasts [72], lactoferrin can inhibit osteoclastic differentiation of RAW264.7 mouse macrophage cell line which induced by the addition of receptor activator for nuclear factor κB ligand (RANKL) [80]. These observations suggest that lactoferrin inhibits bone resorption by reducing osteoclastogenesis. The molecular mechanisms of the inhibitory effect of lactoferrin on osteoclastic differentiation are unknown. The activity is not affect by RAP, suggesting that receptor other than LRP-1 is involved in the process [80].

4.14 Chondrogenic Differentiation

Chondrogenic differentiation is regulated by TGF-β family proteins, including transforming growth factor β (TGF-β) and bone morphogenetic proteins (BMPs) (Fig. 4.3). The members of TGF-β family induce phosphorylation and nuclear translocation of receptor-regulated Smads (R-Smads), which in turn promote cartilage-specific gene expression [81]. In response to TGF-β stimulation, Smad2/3 is translocated to the nucleus, forming a complex with Sox9 to promote the expression

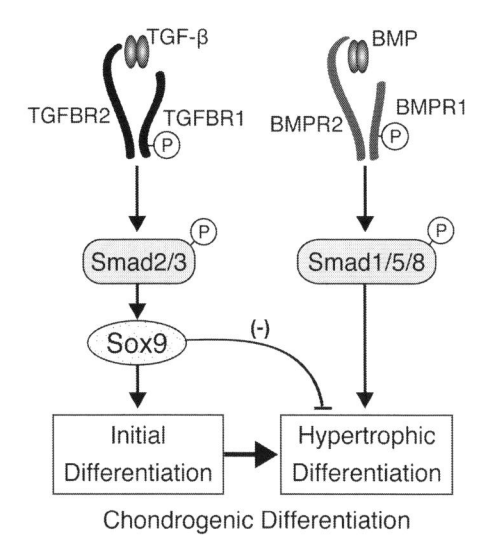

Chondrogenic Differentiation

Fig. 4.3 Regulation of chondrogenic differentiation by receptor-regulated Smads (R-Smads). In response to TGF-β binding, the TGF-β type-II receptor (TGFBR2) phosphorylates and thereby activates the TGF-β type-I receptor (TGFBR1). The activated TGFBR1 phosphorylates Smad2/3. The phosphorylated Smad2/3 migrates to nucleus, leading to transcription of Sox9. Sox9 is a master transcription factor that promotes the initial stage of chondrogenic differentiation. However, Sox9 inhibits the hypertrophic differentiation of chondrocytes. For hypertrophic differentiation, Bone morphogenetic protein2/4 (BMP2/4) activates Smad1/5/8 pathway by recruiting BMP receptor 1 (BMPR1) and BMP receptor 2 (BMPR2)

of genes involved in chondrogenic differentiation [82]. Sox9 is a master transcription factor that is required for the induction of chondrogenic differentiation [83, 84]. On the other hand, Sox9 inhibits the hypertrophic differentiation of chondrocytes [85–87]. Thus, Smad2/3 signals promote the initial stage of chondrogenic differentiation and inhibit hypertrophic differentiation [88]. Smad3-deficient chondrocytes show upregulation of BMP signaling accelerated hypertrophic differentiation [89].

Lactoferrin treatment of differentiating ATDC5 cells (chondroprogenitor cell line) promotes cell proliferation in the initial stage of the differentiation process [90]. The lactoferrin-enhanced proliferation is also observed in primary culture chondrocytes [72, 91]. The promoting effect of lactoferrin is accompanied by elevation of type-II collagen expression [90]. However, lactoferrin treatment inhibits hypertrophic differentiation, characterized by suppression of ALP activity, aggrecan synthesis and N-cadherin expression [90]. This inhibitory effect is accompanied by sustained Sox9 expression, as well as increased Smad2/3 expression and phosphorylation, suggesting that lactoferrin regulates chondrogenic differentiation by up-regulating the Smad2/3-Sox9 signaling pathway.

Accumulating data suggests that lactoferrin can activate Sox9-Smad2/3 signaling pathway. Lactoferrin increases Sox9 expression in C2C12 cells [92]. Lactoferrin treatment enhances nuclear Smad2 translocation in chondrocytes [91] and in Hela-cells [93].

LRP-1 is highly expressed in cartilage tissue and chondrocytes, and plays an important role in the proliferation and differentiation of chondrocytes [91, 94, 95]. However, whether LRP-1 participates in lactoferrin-induced signaling is not known. Lactoferrin activates p38 MAPK and the ERK1/2 in human articular chondrocytes and promote matrix metalloprotease (MMP) expression, suggesting that effect of lactoferrin-enhanced signaling is not restricted in Smad signaling pathway [91].

4.15   Adipogenesis

Oral administration of bovine lactoferrin decreases visceral fat accumulation [96]. The antiadipogenic action of lactoferrin on pre-adipocytes is confirmed by *in vitro* study. Lactoferrin inhibits the adipogenic differentiation of pluripotent mesenchymal cells [97]. The inhibitory effect of lactoferrin on adipogenic differentiation is accompanied by decreased gene expression of CCAAT/enhancer binding protein (C/EBP) α and peroxisome proliferator-activated receptor (PPARγ) [97]. Lactoferrin reduces lipid accumulation in pre-adipocytes derived from the mesenteric fat tissue of rats [98]. Trypsin-treated lactoferrin continues to show anti-adipogenic action, whereas pepsin-treated lactoferrin abrogates the activity [98]. During the adipogenic differentiation of 3 T3-L1 cells, lactoferrin inhibits the expression of lipogenic markers, including fatty acid synthase, acetyl-coenzyme A carboxylase-α and PPAR-γ. Lactoferrin promotes insulin-induced Akt phosphorylation (Ser473) and AMP-activated protein kinase (AMPK) phosphorylation (Thr172) in 3T3-L1 cell line [99].

4.16   Lactoferrin Receptor in Brain Capillary Endothelial Cells

Blood brain barrier (BBB) is localized at the single layer of brain capillary endothelial cells (BCECs). Differentiated bovine BCECs have high affinity (Kd 40 nM) and low affinity (Kd 2 μM) binding site for lactoferrin [15]. Lactoferrin can cross BCECs through receptor-mediated endocytosis and subsequent transcytosis. The translocation of lactoferrin was inhibited by RAP. LRP is responsible for transcytosis of lactoferrin across BCECs [15]. Conjugation of poly (ethyleneglycol)-poly (lactide) nanoparticles (PEG-PLA) with lactoferrin facilitates the translocation of the nanoparticles across BBB [100].

4.17   Platelet Lactoferrin Receptor

Lactoferrin receptor is identified in non-activated human platelets. The KDRS peptide which corresponds to 39–42 of human lactoferrin is identified as platelet lactoferrin binding site [101]. This peptide was identified as inhibitor of platelet aggregation [102].

4.18 Lactoferrin Receptor in Epithelial Cells

Lactoferrin binding activity has been identified in some epithelial cell lines [103]. Lactoferrin interacts with MAC-T bovine mammary epithelial cells [104]. Binding of lactoferrin to MAC-T cells may be associated with the initial events that inhibit MAC-T cell proliferation.

4.19 Lactoferrin Receptor in Respiratory Epithelial Cells

BEAS-2B cells, human bronchial epithelial cells have high-affinity binding sites for lactoferrin [105]. The lactoferrin receptor activity in BEAS-2B cells is enhanced by metal exposure or vanadium, suggesting that lactoferrin contributes to metal sequestration, and prevents an iron-induced oxidative stress to the respiratory tract. However, the extent of iron saturation is not affects to the lactoferrin-binding to BEAS-2B cells. Interestingly, plasmids mixed with lactoferrin that conjugated with branched polyethylenimine (PEI) are more easily taken up by BEAS-2B cells [106]. However, molecular nature of lactoferrin receptor in BEAS-2B cell is not known.

4.20 Transcription Regulator

One of the intriguing roles of lactoferrin is its role as transcription factor. Human lactoferrin has a short cluster of four arginine residues at the N-terminal. This positively charged region is responsible for lactoferrin interaction with heparin, LPS, lysozyme and DNA [107]. Moreover this region acts as nuclear localization signal (NLS) that allows lactoferrin translocation across plasma membrane and nuclear translocation [108, 109]. According to an DNase protection assay, lactoferrin has an affinity for specific DNA sequences such as GCCACTT(G/A)C, TAGA(A/G) GATCAAA and ACTACAGTCTACA [110]. These consensus sequences act as promoter for gene expression in response to human lactoferrin treatment, suggesting the role of lactoferrin as a transcription regulator. Human delta-lactoferrin is known as a transcription factor that enhances Skp1 (S-phase kinase-associated protein) gene expression [111].

References

1. Baker EN, Baker HM (2005) Molecular structure, binding properties and dynamics of lactoferrin. Cell Mol Life Sci 62(22):2531–2539
2. Legrand D, Elass E, Carpentier M, Mazurier J (2006) Interactions of lactoferrin with cells involved in immune function. Biochem Cell Biol 84(3):282–290
3. Suzuki YA, Lopez V, Lonnerdal B (2005) Mammalian lactoferrin receptors: structure and function. Cell Mol Life Sci 62(22):2560–2575

4. Hu WL, Mazurier J, Montreuil J, Spik G (1990) Isolation and partial characterization of a lactotransferrin receptor from mouse intestinal brush border. Biochemistry 29(2):535–541
5. Suzuki YA, Lonnerdal B (2004) Baculovirus expression of mouse lactoferrin receptor and tissue distribution in the mouse. Biometals 17(3):301–309
6. Kawakami H, Lonnerdal B (1991) Isolation and function of a receptor for human lactoferrin in human fetal intestinal brush-border membranes. Am J Physiol 261(5 Pt 1):G841–G846
7. Tsuji S, Uehori J, Matsumoto M, Suzuki Y et al (2001) Human intelectin is a novel soluble lectin that recognizes galactofuranose in carbohydrate chains of bacterial cell wall. J Biol Chem 276(26):23456–23463
8. Suzuki YA, Shin K, Lonnerdal B (2001) Molecular cloning and functional expression of a human intestinal lactoferrin receptor. Biochemistry 40(51):15771–15779
9. Willnow TE, Goldstein JL, Orth K, Brown MS et al (1992) Low density lipoprotein receptor-related protein and gp330 bind similar ligands, including plasminogen activator-inhibitor complexes and lactoferrin, an inhibitor of chylomicron remnant clearance. J Biol Chem 267(36):26172–26180
10. Herz J, Strickland DK (2001) LRP: a multifunctional scavenger and signaling receptor. J Clin Invest 108(6):779–784
11. May P, Woldt E, Matz RL, Boucher P (2007) The LDL receptor-related protein (LRP) family: an old family of proteins with new physiological functions. Ann Med 39(3):219–228
12. Fisher CE, Howie SE (2006) The role of megalin (LRP-2/Gp330) during development. Dev Biol 296(2):279–297
13. Lillis AP, Greenlee MC, Mikhailenko I, Pizzo SV et al (2008) Murine low-density lipoprotein receptor-related protein 1 (LRP) is required for phagocytosis of targets bearing LRP ligands but is not required for C1q-triggered enhancement of phagocytosis. J Immunol 181(1):364–373
14. Neels JG, van Den Berg BM, Lookene A, Olivecrona G et al (1999) The second and fourth cluster of class A cysteine-rich repeats of the low density lipoprotein receptor-related protein share ligand-binding properties. J Biol Chem 274(44):31305–31311
15. Fillebeen C, Descamps L, Dehouck MP, Fenart L et al (1999) Receptor-mediated transcytosis of lactoferrin through the blood-brain barrier. J Biol Chem 274(11):7011–7017
16. Takayama Y, Takahashi H, Mizumachi K, Takezawa T (2003) Low density lipoprotein receptor-related protein (LRP) is required for lactoferrin-enhanced collagen gel contractile activity of human fibroblasts. J Biol Chem 278(24):22112–22118
17. Grey A, Banovic T, Zhu Q, Watson M et al (2004) The low-density lipoprotein receptor-related protein 1 is a mitogenic receptor for lactoferrin in osteoblastic cells. Mol Endocrinol 18(9):2268–2278
18. Tang L, Wu JJ, Ma Q, Cui T et al (2010) Human lactoferrin stimulates skin keratinocyte function and wound re-epithelialization. Br J Dermatol 163(1):38–47
19. Prieels JP, Pizzo SV, Glasgow LR, Paulson JC et al (1978) Hepatic receptor that specifically binds oligosaccharides containing fucosyl alpha1 leads to 3 N-acetylglucosamine linkages. Proc Natl Acad Sci USA 75(5):2215–2219
20. Bennett RM, Kokocinski T (1979) Lactoferrin turnover in man. Clin Sci (Lond) 57(5):453–460
21. Ziere GJ, Bijsterbosch MK, van Berkel TJ (1993) Removal of 14 N-terminal amino acids of lactoferrin enhances its affinity for parenchymal liver cells and potentiates the inhibition of beta- very low density lipoprotein binding. J Biol Chem 268(36):27069–27075
22. Bennatt DJ, McAbee DD (1997) Identification and isolation of a 45-kDa calcium-dependent lactoferrin receptor from rat hepatocytes. Biochemistry 36(27):8359–8366
23. McAbee DD, Bennatt DJ, Ling YY (1998) Identification and analysis of a CA(2+)-dependent lactoferrin receptor in rat liver. Lactoferrin binds to the asialoglycoprotein receptor in a galactose-independent manner. Adv Exp Med Biol 443:113–121
24. Ginsty H, Amalric F, Bouvet P (1998) Nucleolin functions in the first step of ribosomal RNA processing. EMBO J 17(5):1476–1486

25. Srivastava M, Pollard HB (1999) Molecular dissection of nucleolin's role in growth and cell proliferation: new insights. FASEB J 13(14):1911–1922
26. Kleinman HK, Weeks BS, Cannon FB, Sweeney TM et al (1991) Identification of a 110-kDa nonintegrin cell surface laminin-binding protein which recognizes an A chain neurite-promoting peptide. Arch Biochem Biophys 290(2):320–325
27. Take M, Tsutsui J, Obama H, Ozawa M et al (1994) Identification of nucleolin as a binding protein for midkine (MK) and heparin-binding growth associated molecule (HB-GAM). J Biochem 116(5):1063–1068
28. Said EA, Krust B, Nisole S, Svab J et al (2002) The anti-HIV cytokine midkine binds the cell surface-expressed nucleolin as a low affinity receptor. J Biol Chem 277(40):37492–37502
29. Larrucea S, Gonzalez-Rubio C, Cambronero R, Ballou B et al (1998) Cellular adhesion mediated by factor J, a complement inhibitor. Evidence for nucleolin involvement. J Biol Chem 273(48):31718–31725
30. Legrand D, Vigie K, Said EA, Elass E et al (2004) Surface nucleolin participates in both the binding and endocytosis of lactoferrin in target cells. Eur J Biochem 271(2):303–317
31. Legrand D, van Berkel PH, Salmon V, van Veen HA et al (1997) The N-terminal Arg2, Arg3 and Arg4 of human lactoferrin interact with sulphated molecules but not with the receptor present on Jurkat human lymphoblastic T-cells. Biochem J 327(Pt 3):841–846
32. Mazurier J, Legrand D, Hu WL, Montreuil J et al (1989) Expression of human lactotransferrin receptors in phytohemagglutinin-stimulated human peripheral blood lymphocytes. Isolation of the receptors by antiligand-affinity chromatography. Eur J Biochem 179(2):481–487
33. Leveugle B, Mazurier J, Legrand D, Mazurier C et al (1993) Lactotransferrin binding to its platelet receptor inhibits platelet aggregation. Eur J Biochem 213(3):1205–1211
34. Van Snick JL, Masson PL (1976) The binding of human lactoferrin to mouse peritoneal cells. J Exp Med 144(6):1568–1580
35. Roseanu A, Chelu F, Trif M, Motas C et al (2000) Inhibition of binding of lactoferrin to the human promonocyte cell line THP-1 by heparin: the role of cell surface sulphated molecules. Biochim Biophys Acta 1475(1):35–38
36. Eda S, Kikugawa K, Beppu M (1997) Characterization of lactoferrin-binding proteins of human macrophage membrane: multiple species of lactoferrin-binding proteins with polylactosamine-binding ability. Biol Pharm Bull 20(2):127–133
37. Haversen L, Ohlsson BG, Hahn-Zoric M, Hanson LA et al (2002) Lactoferrin down-regulates the LPS-induced cytokine production in monocytic cells via NF-kappa B. Cell Immunol 220(2):83–95
38. Mattsby-Baltzer I, Roseanu A, Motas C, Elverfors J et al (1996) Lactoferrin or a fragment thereof inhibits the endotoxin-induced interleukin-6 response in human monocytic cells. Pediatr Res 40(2):257–262
39. Elass-Rochard E, Legrand D, Salmon V, Roseanu A et al (1998) Lactoferrin inhibits the endotoxin interaction with CD14 by competition with the lipopolysaccharide-binding protein. Infect Immun 66(2):486–491
40. Ando K, Hasegawa K, Shindo K, Furusawa T et al (2010) Human lactoferrin activates NF-kappaB through the Toll-like receptor 4 pathway while it interferes with the lipopolysaccharide-stimulated TLR4 signaling. FEBS J 277(9):2051–2066
41. Curran CS, Demick KP, Mansfield JM (2006) Lactoferrin activates macrophages via TLR4-dependent and -independent signaling pathways. Cell Immunol 242(1):23–30
42. Pluddemann A, Neyen C, Gordon S (2007) Macrophage scavenger receptors and host-derived ligands. Methods 43(3):207–217
43. Hirano K, Miki Y, Hirai Y, Sato R et al (2005) A multifunctional shuttling protein nucleolin is a macrophage receptor for apoptotic cells. J Biol Chem 280(47):39284–39293
44. Zimecki M, Kocieba M, Kruzel M (2002) Immunoregulatory activities of lactoferrin in the delayed type hypersensitivity in mice are mediated by a receptor with affinity to mannose. Immunobiology 205(1):120–131

45. Groot F, Geijtenbeek TB, Sanders RW, Baldwin CE et al (2005) Lactoferrin prevents dendritic cell-mediated human immunodeficiency virus type 1 transmission by blocking the DC-SIGN–gp120 interaction. J Virol 79(5):3009–3015

46. Naarding MA, Ludwig IS, Groot F, Berkhout B et al (2005) Lewis X component in human milk binds DC-SIGN and inhibits HIV-1 transfer to CD4+ T lymphocytes. J Clin Invest 115(11):3256–3264

47. Bennett RM, Davis J (1981) Lactoferrin binding to human peripheral blood cells: an interaction with a B-enriched population of lymphocytes and a subpopulation of adherent mononuclear cells. J Immunol 127(3):1211–1216

48. Zimecki M, Mazurier J, Spik G, Kapp JA (1995) Human lactoferrin induces phenotypic and functional changes in murine splenic B cells. Immunology 86(1):122–127

49. Mincheva-Nilsson L, Hammarstrom S, Hammarstrom ML (1997) Activated human gamma delta T lymphocytes express functional lactoferrin receptors. Scand J Immunol 46(6):609–618

50. Bi BY, Liu JL, Legrand D, Roche AC et al (1996) Internalization of human lactoferrin by the Jurkat human lymphoblastic T-cell line. Eur J Cell Biol 69(3):288–296

51. Dhennin-Duthille I, Masson M, Damiens E, Fillebeen C et al (2000) Lactoferrin upregulates the expression of CD4 antigen through the stimulation of the mitogen-activated protein kinase in the human lymphoblastic T Jurkat cell line. J Cell Biochem 79(4):583–593

52. Frydecka I, Zimecki M, Bocko D, Kosmaczewska A et al (2002) Lactoferrin-induced up-regulation of zeta (zeta) chain expression in peripheral blood T lymphocytes from cervical cancer patients. Anticancer Res 22(3):1897–1901

53. Damiens E, El Yazidi I, Mazurier J, Duthille I et al (1999) Lactoferrin inhibits G1 cyclin-dependent kinases during growth arrest of human breast carcinoma cells. J Cell Biochem 74(3):486–498

54. Son HJ, Lee SH, Choi SY (2006) Human lactoferrin controls the level of retinoblastoma protein and its activity. Biochem Cell Biol 84(3):345–350

55. Oh SM, Pyo CW, Kim Y, Choi SY (2004) Neutrophil lactoferrin upregulates the human p53 gene through induction of NF-kappaB activation cascade. Oncogene 23(50):8282–8291

56. Zhou Y, Zeng Z, Zhang W, Xiong W et al (2008) Lactotransferrin: a candidate tumor suppressor-Deficient expression in human nasopharyngeal carcinoma and inhibition of NPC cell proliferation by modulating the mitogen-activated protein kinase pathway. Int J Cancer 123(9):2065–2072

57. Xiao Y, Monitto CL, Minhas KM, Sidransky D (2004) Lactoferrin down-regulates G1 cyclin-dependent kinases during growth arrest of head and neck cancer cells. Clin Cancer Res 10(24):8683–8686

58. Xu XX, Jiang HR, Li HB, Zhang TN et al (2010) Apoptosis of stomach cancer cell SGC-7901 and regulation of Akt signaling way induced by bovine lactoferrin. J Dairy Sci 93(6):2344–2350

59. Lee SH, Park SW, Pyo CW, Yoo NK et al (2009) Requirement of the JNK-associated Bcl-2 pathway for human lactoferrin-induced apoptosis in the Jurkat leukemia T cell line. Biochimie 91(1):102–108

60. Lee SH, Hwang HM, Pyo CW, Hahm DH et al (2010) E2F1-directed activation of Bcl-2 is correlated with lactoferrin-induced apoptosis in Jurkat leukemia T lymphocytes. Biometals 23(3):507–514

61. Furlong SJ, Mader JS, Hoskin DW (2010) Bovine lactoferricin induces caspase-independent apoptosis in human B-lymphoma cells and extends the survival of immune-deficient mice bearing B-lymphoma xenografts. Exp Mol Pathol 88(3):371–375

62. Lee SH, Pyo CW, Hahm DH, Kim J et al (2009) Iron-saturated lactoferrin stimulates cell cycle progression through PI3K/Akt pathway. Mol Cells 28(1):37–42

63. Breton M, Mariller C, Benaissa M, Caillaux K et al (2004) Expression of delta-lactoferrin induces cell cycle arrest. Biometals 17(3):325–329

64. Norrby K (2004) Human apo-lactoferrin enhances angiogenesis mediated by vascular endothelial growth factor A in vivo. J Vasc Res 41(4):293–304

65. Kim CW, Son KN, Choi SY, Kim J (2006) Human lactoferrin upregulates expression of KDR/Flk-1 and stimulates VEGF-A-mediated endothelial cell proliferation and migration. FEBS Lett 580(18):4332–4336

66. Norrby K, Mattsby-Baltzer I, Innocenti M, Tuneberg S (2001) Orally administered bovine lactoferrin systemically inhibits VEGF(165)-mediated angiogenesis in the rat. Int J Cancer 91(2):236–240

67. Shimamura M, Yamamoto Y, Ashino H, Oikawa T et al (2004) Bovine lactoferrin inhibits tumor-induced angiogenesis. Int J Cancer 111(1):111–116

68. Mader JS, Smyth D, Marshall J, Hoskin DW (2006) Bovine lactoferricin inhibits basic fibroblast growth factor- and vascular endothelial growth factor165-induced angiogenesis by competing for heparin-like binding sites on endothelial cells. Am J Pathol 169(5):1753–1766

69. Takayama Y, Mizumachi K (2001) Effects of lactoferrin on collagen gel contractile activity and myosin light chain phosphorylation in human fibroblasts. FEBS Lett 508(1):111–116

70. Tang L, Cui T, Wu JJ, Liu-Mares W et al (2010) A rice-derived recombinant human lactoferrin stimulates fibroblast proliferation, migration, and sustains cell survival. Wound Repair Regen 18(1):123–131

71. Birkenmeier G, Heidrich K, Glaser C, Handschug K et al (1998) Different expression of the alpha2-macroglobulin receptor/low-density lipoprotein receptor-related protein in human keratinocytes and fibroblasts. Arch Dermatol Res 290(10):561–568

72. Cornish J, Callon KE, Naot D, Palmano KP et al (2004) Lactoferrin is a potent regulator of bone cell activity and increases bone formation in vivo. Endocrinology 145(9):4366–4374

73. Takayama Y, Mizumachi K (2008) Effect of bovine lactoferrin on extracellular matrix calcification by human osteoblast-like cells. Biosci Biotechnol Biochem 72(1):226–230

74. Takayama Y, Mizumachi K (2009) Effect of lactoferrin-embedded collagen membrane on osteogenic differentiation of human osteoblast-like cells. J Biosci Bioeng 107(2):191–195

75. Herz J, Goldstein JL, Strickland DK, Ho YK et al (1991) 39-kDa protein modulates binding of ligands to low density lipoprotein receptor-related protein/alpha 2-macroglobulin receptor. J Biol Chem 266(31):21232–21238

76. Willnow TE, Sheng Z, Ishibashi S, Herz J (1994) Inhibition of hepatic chylomicron remnant uptake by gene transfer of a receptor antagonist. Science 264(5164):1471–1474

77. Grey A, Zhu Q, Watson M, Callon K et al (2006) Lactoferrin potently inhibits osteoblast apoptosis, via an LRP1-independent pathway. Mol Cell Endocrinol 251(1–2):96–102

78. Nakajima KI, Kanno Y, Nakamura M, Gao XD et al. (2011) Bovine milk lactoferrin induces synthesis of the angiogenic factors VEGF and FGF2 in osteoblasts via the p44/p42 MAP kinase pathway. Biometals 24(5):847–856

79. Lorget F, Clough J, Oliveira M, Daury MC et al (2002) Lactoferrin reduces in vitro osteoclast differentiation and resorbing activity. Biochem Biophys Res Commun 296(2):261–266

80. Cornish J, Naot D (2010) Lactoferrin as an effector molecule in the skeleton. Biometals 23(3):425–430

81. Li TF, O'Keefe RJ, Chen D (2005) TGF-beta signaling in chondrocytes. Front Biosci 10:681–688

82. Furumatsu T, Tsuda M, Taniguchi N, Tajima Y et al (2005) Smad3 induces chondrogenesis through the activation of SOX9 via CREB-binding protein/p300 recruitment. J Biol Chem 280(9):8343–8350

83. Akiyama H (2008) Control of chondrogenesis by the transcription factor Sox9. Mod Rheumatol 18(3):213–219

84. Provot S, Schipani E (2005) Molecular mechanisms of endochondral bone development. Biochem Biophys Res Commun 328(3):658–665

85. Ikeda T, Kawaguchi H, Kamekura S, Ogata N et al (2005) Distinct roles of Sox5, Sox6, and Sox9 in different stages of chondrogenic differentiation. J Bone Miner Metab 23(5):337–340

86. Huang W, Chung UI, Kronenberg HM, de Crombrugghe B (2001) The chondrogenic transcription factor Sox9 is a target of signaling by the parathyroid hormone-related peptide in the growth plate of endochondral bones. Proc Natl Acad Sci USA 98(1):160–165

87. Yamashiro T, Wang XP, Li Z, Oya S et al (2004) Possible roles of Runx1 and Sox9 in incipient intramembranous ossification. J Bone Miner Res 19(10):1671–1677
88. Yang X, Chen L, Xu X, Li C et al (2001) TGF-beta/Smad3 signals repress chondrocyte hypertrophic differentiation and are required for maintaining articular cartilage. J Cell Biol 153(1):35–46
89. Li TF, Darowish M, Zuscik MJ, Chen D et al (2006) Smad3-deficient chondrocytes have enhanced BMP signaling and accelerated differentiation. J Bone Miner Res 21(1):4–16
90. Takayama Y, Mizumachi K (2010) Inhibitory effect of lactoferrin on hypertrophic differentiation of ATDC5 mouse chondroprogenitor cells. Biometals 23(3):477–484
91. Brandl N, Zemann A, Kaupe I, Marlovits S et al (2010) Signal transduction and metabolism in chondrocytes is modulated by lactoferrin. Osteoarthritis Cartilage 18(1):117–125
92. Yagi M, Suzuki N, Takayama T, Arisue M et al (2009) Effects of lactoferrin on the differentiation of pluripotent mesenchymal cells. Cell Biol Int 33(3):283–289
93. Zemann N, Klein P, Wetzel E, Huettinger F et al (2010) Lactoferrin induces growth arrest and nuclear accumulation of Smad-2 in HeLa cells. Biochimie 92(7):880–884
94. Kawata K, Kubota S, Eguchi T, Moritani NH et al (2010) Role of the low-density lipoprotein receptor-related protein-1 in regulation of chondrocyte differentiation. J Cell Physiol 222(1):138–148
95. Kawata K, Eguchi T, Kubota S, Kawaki H et al (2006) Possible role of LRP1, a CCN2 receptor, in chondrocytes. Biochem Biophys Res Commun 345(2):552–559
96. Ono T, Murakoshi M, Suzuki N, Iida N et al (2010) Potent anti-obesity effect of enteric-coated lactoferrin: decrease in visceral fat accumulation in Japanese men and women with abdominal obesity after 8-week administration of enteric-coated lactoferrin tablets. Br J Nutr 104(11):1688–1695
97. Yagi M, Suzuki N, Takayama T, Arisue M et al (2008) Lactoferrin suppress the adipogenic differentiation of MC3T3-G2/PA6 cells. J Oral Sci 50(4):419–425
98. Ono T, Morishita S, Fujisaki C, Ohdera M et al (2011) Effects of pepsin and trypsin on the anti-adipogenic action of lactoferrin against pre-adipocytes derived from rat mesenteric fat. Br J Nutr 105(2):200–211
99. Moreno-Navarrete JM, Ortega FJ, Ricart W, Fernandez-Real JM (2009) Lactoferrin increases (172Thr)AMPK phosphorylation and insulin-induced (p473Ser)AKT while impairing adipocyte differentiation. Int J Obes (Lond) 33(9):991–1000
100. Hu K, Li J, Shen Y, Lu W et al (2009) Lactoferrin-conjugated PEG-PLA nanoparticles with improved brain delivery: in vitro and in vivo evaluations. J Control Release 134(1):55–61
101. Maneva A, Taleva B, Manev V, Sirakov L (1993) Lactoferrin binding to human platelets. Int J Biochem 25(5):707–712
102. Mazoyer E, Levy-Toledano S, Rendu F, Hermant L et al (1990) KRDS, a new peptide derived from human lactotransferrin, inhibits platelet aggregation and release reaction. Eur J Biochem 194(1):43–49
103. Rochard E, Legrand D, Lecocq M, Hamelin R (1992) Characterization of lactotransferrin receptor in epithelial cell lines from non-malignant human breast, benign mastopathies and breast carcinomas. Anticancer Res 12(6B):2047–2051
104. Rejman JJ, Turner JD, Oliver SP (1994) Characterization of lactoferrin binding to the MAC-T bovine mammary epithelial cell line using a biotin-avidin technique. Int J Biochem 26(2):201–206
105. Ghio AJ, Carter JD, Dailey LA, Devlin RB et al (1999) Respiratory epithelial cells demonstrate lactoferrin receptors that increase after metal exposure. Am J Physiol 276(6 Pt 1):L933–L940
106. Elfinger M, Maucksch C, Rudolph C (2007) Characterization of lactoferrin as a targeting ligand for nonviral gene delivery to airway epithelial cells. Biomaterials 28(23):3448–3455
107. van Berkel PH, Geerts ME, van Veen HA, Mericskay M et al (1997) N-terminal stretch Arg2, Arg3, Arg4 and Arg5 of human lactoferrin is essential for binding to heparin, bacterial lipopolysaccharide, human lysozyme and DNA. Biochem J 328(Pt 1):145–151

108. Penco S, Scarfì S, Giovine M, Damonte G et al (2001) Identification of an import signal for, and the nuclear localization of, human lactoferrin. Biotechnol Appl Biochem 34(Pt 3):151–159
109. Garre C, Bianchi-Scarra G, Sirito M, Musso M et al (1992) Lactoferrin binding sites and nuclear localization in K562(S) cells. J Cell Physiol 153(3):477–482
110. He J, Furmanski P (1995) Sequence specificity and transcriptional activation in the binding of lactoferrin to DNA. Nature 373(6516):721–724
111. Mariller C, Benaissa M, Hardiville S, Breton M et al (2007) Human delta-lactoferrin is a transcription factor that enhances Skp1 (S-phase kinase-associated protein) gene expression. FEBS J 274(8):2038–2053

Chapter 5
Effects of Lactoferrin on Skin Wound Healing

Abstract The skin wound healing is a complex biological process that requires the regulation of different cell types. Lactoferrin is a metal-binding glycoprotein secreted from glandular epithelial cells and neutrophils. The topical administration of lactoferrin enhances the rate of skin wound closure in normal and diabetic mice. The promoting effect of lactoferrin on skin wound healing is partially dues to its immuno-modulating properties. Lactoferrin promotes the initial stage of inflammatory phase by increasing the production of pro-inflammatory cytokines and infiltration of immune cells into wounded area. On the other hand, lactoferrin is likely to serve as anti-inflammatory agent that neutralizes overabundant immune response. Moreover, lactoferrin is thought to promote both the granulation tissue formation and reepithelialization by enhancing the proliferation and migration of fibroblasts and keratinocytes. The synthesis of extracellular matrix components is also enhanced by lactoferrin. In an *in vitro* model of wound contraction, lactoferrin stimulates fibroblast-mediated collagen gel contraction. Lactoferrin is known as its anti-bacterial, anti-viral activities which may contribute to the healing of diabetic ulcers. These lines of observations indicate that lactoferrin can support the multiple biological processes involved in wound healing. Based on these findings, lactoferrin could be used in patients with diabetic and other types of ulcers.

Keywords Diabetic wounds • Inflammation • Reepithelialization • Wound contraction

5.1 Introduction

The skin wound healing is a complex and dynamic biological process that requires the regulation of different cells, including immune cells, keratinocytes, fibroblasts, and endothelial cells [1–3]. It consists of hemostasis, inflammation, granulation tissue formation, angiogenesis, reepithelialization and wound resolution. Wound healing begins

Fig. 5.1 Effects of lactoferrin on normal wound healing. Lactoferrin can promote the sequential events of wound healing, such as inflammation, granulation tissue formation and inflammation

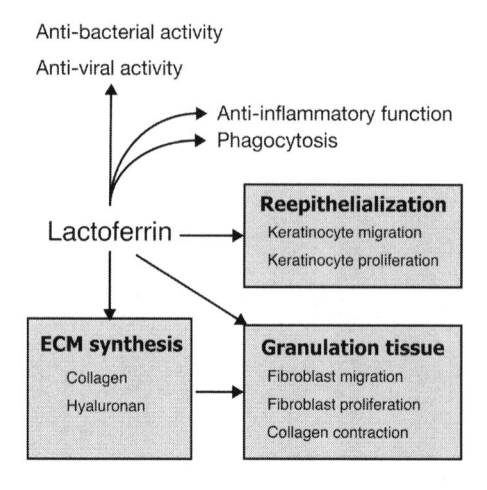

with hemostasis and is followed by the inflammatory phase. The inflammatory phase is prerequisite for healthy wound repair to kill bacteria and remove cell debris and extracellular matrix fragments [2]. This process begins with infiltration of neutrophils and macrophages into wounded area [4]. Lactoferrin is a metal-binding glycoprotein belonging to the transferring family. It is synthesized by glandular epithelial cells and secreted into body fluids such as saliva, tears and mucosal secretions. Lactoferrin participates in host defenses, due to its anti-bacterial, anti-viral and anti-fungal activities. One of the novel activities described for lactoferrin is its regulatory function in wound healing (Fig. 5.1). The results of animal studies indicate that the topical administration of lactoferrin enhances the rate of skin wound closure in normal and diabetic mice. Lactoferrin can act as both positive and negative regulator of innate and adaptive immune cells [5–8]. Lactoferrin is known as a major component of the secondary granules of polymorphonuclear neutrophils (PMNs) and is released into plasma during infection or inflammation. Lactoferrin promotes early inflammatory phase of wound healing by increasing the production of the pro-inflammatory cytokines, and thereby augments the infiltration of neutrophils and macrophages into wounded area. They kill invading microorganisms by producing reactive oxygen species (ROS) and clear up matrix and cell debris by phagocytosis [2, 9]. On the other hand, prolonged or excess inflammation, characterized by overabundance of neutrophils and macrophages in the wound area, leads to non-healing chronic wound [2]. Most non-healing chronic wounds fail to progress through the normal phase of wound repair and remain in an inflammation state [10]. Interestingly, lactoferrin appears to serve as exhibit anti-inflammatory activity to neutralize an overabundant immune response.

In addition, recent *in vivo* studies indicate that lactoferrin directly increases the proliferation, and migration of fibroblasts. Fibroblasts are predominant cells in dermis and responsible for generation of granulation tissue in wound healing process. The synthesis of extracellular matrix components, such as collagen and hyaluronan, is enhanced by lactoferrin. In an *in vitro* model of wound contraction, lactoferrin promotes fibroblast-mediated collagen gel contraction. The proliferation and migration of keratinocyte is essential for reepithelialization of wound healing [1, 11]. The promoting effect of lactoferrin on keratinocyte proliferation and migration has been reported. In addition, *in vivo* animal studies confirm the promoting effect of lactoferrin on wound healing. There data implicates lactoferrin contributes to its efficacy in accelerating normal and chronic wound repair.

5.2 Cell Proliferation and Migration

Cell proliferation and migration is critical for wound healing, especially granulation tissue formation and reepithelialization. In response to injury, fibroblasts in proximity of the wound begin to proliferate and migrate into the wound area. The effect of lactoferrin on cell proliferation and migration is dependent on cell types. Lactoferrin inhibits the migration of gastrointestinal cells [12]. Bovine lactoferrin inhibits proliferation of human corneal limbal epithelial (HCLE) cells. In addition, lactoferrin antagonizes the migration of HCLE cells that induced by platelet derived growth factor (PDGF) or IL-6 in combination with fibronectin [13]. The promoting effect of lactoferrin on wound healing is based on its ability to promote the migration of fibroblasts, as shown in a wound healing assay [14] and in a trans-well migration assay [15, 16]. In addition, human recombinant lactoferrin has an ability to enhance fibroblast proliferation [16]. Lactoferrin has positive synergistic effects with fibroblast growth factor-2 (FGF-2/bFGF) on fibroblast proliferation, and antagonizes the inhibitory effect of transforming growth factor β (TGF-β) on fibroblast proliferation. Moreover, the promoting effects of human recombinant lactoferrin on proliferation and migration are also observed in human keratinocytes, suggesting that lactoferrin has an ability to promote reepithelialization [17]. Keratinocytes is major types of cells in epidermis and represent the body's first line of defense from the outside environment. After mechanical injury, keratinocytes start proliferating and migrating into wounded area to cover the wounded area [1, 11, 18]. These results are consistent with the study using animal model. Recombinant human lactoferrin can promote the reepithelialization of porcine second-degree burn wound [17].

5.3 Collagen Gel Contraction by Fibroblasts

A role for lactoferrin in wound healing was proposed based on the observation that lactoferrin promotes the fibroblast-mediated collagen gel contraction [14]. Fibroblasts cultured in a three-dimensional type-I collagen gel are able to reorganize the surrounding collagen gel matrix into a more denser and more compact structure. This phenomenon, referred to as collagen gel contraction, is used as *in vitro* model of wound contraction. The extent of collagen gel contraction appears to reflect the motility of the cells in the collagen gel [19, 20]. Wound contraction contributes for the reduction of the wound size and therefore to shorten the healing period. It involves

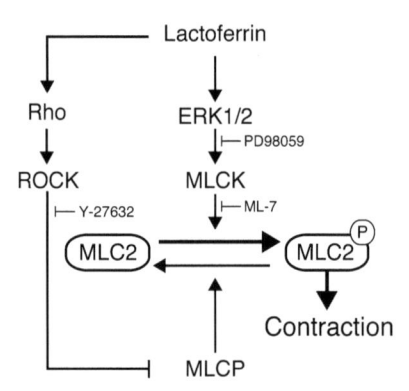

Fig. 5.2 Lactoferrin promotes myosin light chain 2 (MLC2) phosphorylation, and collagen gel contraction by two pathways. Lactoferrin activates myosin light chain kinase (MLCK) via ERK1/2-dependent mechanism, thereby facilitates MLC2 phosphorylation. Study with protein kinase inhibitors suggests that lactoferrin can promote MLC2 phosphorylation through activation of Rho-kinase (ROCK) and the resulting inactivation of myosin light chain phosphatase (MLCP). PD98059 is an ERK1/2 pathway inhibitor. ML-7 is a MLCK inhibitor. Y-27632 is a specific antagonist of ROCK

numerous actions, including cell adhesion to the collagen fibers, cell migration through the collagen matrix. Certain cytokines and growth factors such as TGF-β, PDGF, and FGF, are known to promote collagen gel contraction by fibroblasts.

Bovine and human lactoferrins promote the collagen gel contraction by WI-38 human fibroblasts in a dose-dependent manner [14]. The effect of apo-lactoferrin and holo-lactoferrin on collagen gel contraction is almost the same. The fragment corresponding to the C-lobe of bovine lactoferrin (amino acids 341–689) has a more prominent effect on collagen gel contractile activity than that of either native bovine lactoferrin or its N-lobe (1–284) [21]. The phosphorylation of myosine light chain 2 (MLC2) at Ser 19 is a critical step in the regulation of the migratory and collagen gel contractile activities of fibroblasts [22], as well as in the regulation of actin-myosin interaction and actin stress fiber formation in smooth muscle cells and non-smooth muscle cells. Lactoferrin may play a role in this process as well, as it increases MLC phosphorylation in human fibroblasts [14]. Both MLC kinase (MLCK) inhibitor ML-7 and Rho kinase (ROK/ROCK) inhibitor Y-27632 inhibit lactoferrin induced MLC phosphorylation and collagen gel contraction, suggesting that two distinct pathways, MLCK-dependent and ROCK-dependent, are involved in the promoting effects of lactoferrin (Fig. 5.2) [14].

MLCK is a Ca^{2+} calmodulin-dependent serine/threonine kinase. It is a critical regulator of MLC phosphorylation in fibroblasts and myo-fibroblasts as well as in smooth muscle cells [23]. MLCK is involved in focal adhesion turnover and membrane protrusion at the front of migrating cells [24, 25]. The kinase activity of MLCK is regulated by ERK1/2 (p42/44 mitogen activated protein kinase). ERK1/2 is activated in the process of mitogen-induced cell migration and wound healing [26]. ERK1/2 phosphorylates MLCK and increases its kinase activity [27, 28]. The lactoferrin-treatment of fibroblast results in activation ERK1/2 and, thereby MLCK

[29]. Lactoferrin-induced MLCK activation and collagen gel contraction are inhibited by PD98059, an ERK pathway inhibitor [14, 29], suggesting that the lactoferrin-enhanced MLCK activation is dependent on ERK1/2. These results imply that pathway dependent on ERK1/2 and MLCK participates in the lactoferrin-enhanced collagen gel contraction in human fibroblasts (Fig. 5.2).

The small G protein Rho participates in collagen gel contraction and migration of fibroblasts. ROCK is one of the Rho GTPase-sensitive serine/threonine kinase, and is another important regulator of MLC phosphorylation and the contractile activity of fibroblasts [30]. ROCK-induced phosphorylation of MLC phosphatase (MLCP) results in the elevation of MLC phosphorylation, thereby promoting the collagen gel contraction force of fibroblasts [31]. In addition, ROCK can directly phosphorylate MLC [23, 32]. Both exoenzyme C3 (a Rho inactivating enzyme) and Y-27632 block the lactoferrin-enhanced collagen gel contraction [14]. As well as MLCK, the ROCK pathway is likely to involved in lactoferrin-enhanced collagen gel contraction by increasing MLC phosphorylation.

Interestingly, the lactoferrin-enhanced proliferation and migration of human dermal fibroblasts are blocked by PD98059 and Y-27632 [16]. Taken together, lactoferrin can act as a growth factor, enhancing the proliferation and migration of fibroblasts by activating common signaling pathway critical for collagen gel contraction.

5.4 Molecular Mechanism of Wound Healing

Lactoferrin exerts its biological effects by interacting with its receptor of targeting cells. Lactoferrin can interact with extracellular domain of low-density lipoprotein receptor related protein-1 (LRP-1) [33, 34]. LRP-1 is an endocytic receptor that belongs to LDL receptor family [35, 36]. In addition to lactoferrin, LRP-1 can bind diverse ligands including apolipoprotein E (Apo E), $\alpha2$ macroglobulin, tissue-type plasminogen activator (tPA), and urokinase-type plasminogen activator (uPA) [37]. LRP-1 is involved in regulation of cell motility or contractility. Inhibition of LRP-1 function prevents smooth muscle cell migration induced by uPA [38]. Hsp90α promotes both epidermal and dermal cell migration through LRP-1-dependent mechanism [39]. Promoting effect of uPA on contraction of smooth muscle cells is dependent on LRP-1 [40]. LRP-1 has been implicated as a component of the receptor complex for midkine, a heparin binding growth factor with collagen contraction promoting activities [41, 42].

LRP-1 is involved in lactoferrin metabolism by mediating endocytic clearance of lactoferrin from the circulation by hepatocytes [43, 44]. Transcytosis of lactoferrin across blood brain barrier is mediated by LRP-1 [45, 46]. The lactoferrin-induced activation of cAMP-dependent protein kinase (PKA) in M21 human melanoma cells is inhibited by receptor-associated protein (RAP) [47]. RAP is a high affinity ligand for LRP-1 and is utilized as a universal competitor of LRP-1 ligands [48]. These lines of observations suggest that LRP-1 acts as a physiological receptor for lactoferrin.

LRP-1 expression is found in human dermal fibroblasts and epidermal keratino-cytes [49]. Lactoferrin specifically binds to WI-38 human fibroblasts [14], and this binding is inhibited with RAP [29]. Furthermore, suppression of LRP-1 expression by antisense-oligonucleotide against LRP-1 reverses the promoting effect of lacto-ferrin on collagen gel contraction by WI-38 human fibroblasts [29]. Lactoferrin-induced ERK1/2 phosphorylation and MLC phosphorylation are also inhibited by the suppression of LRP-1 expression [29]. The mitogenic effect of lactoferrin on fibroblast is dependent on LRP-1 [16, 50]. As well as fibroblasts, the promoting effect of lactoferrin on proliferation and migration of human epithelial keratino-cytes are blocked by PD98059 and Y-27632 [17]. Down regulation of LRP-1 expression antagonizes the promoting effect of lactoferrin on keratinocyte migra-tion [17].

Based on these observations, LRP-1 is regarded as lactoferrin signaling mediator in fibroblasts and keratinocytes to elicit the activation of ERK1/2 and Rho in response to lactoferrin. Lactoferrin can activate ERK1/2 in osteoblasts and T-cells [50, 51]. However, suppression of LRP-1 expression did not inhibit the specific binding of lactoferrin to WI-38 human fibroblasts [29], suggesting that LRP-1 is not the initial binding site for lactoferrin on human fibroblasts. The molecular mecha-nism by which LRP-1 activate the ERK1/2 is not fully understood. The cytoplasmic domain of LRP-1 interacts with cytoplasmic adaptor proteins that are involved in the regulation of ERK1/2 activity, cytoskeletal reorganization, and cell adhesion. These adaptor proteins include Shc, FE65, PSD-95, and c-Jun amino-terminal kinase-interacting proteins (JIPs) [52–54].

5.5   Regulation of Cell Adhesion

As described above, cell adhesion is a critical step for cell migration. The effects of lactoferrin to cell adhesion are dependent on the types of cells. Lactoferrin inhibits the adhesions of fibroblasts and intestinal epithelial cells [55, 56]. Lactoferrin binds to the RGD-containing human extracellular matrix proteins, fibronectin and vit-ronectin. Furthermore, human lactoferrin inhibits cell adhesion to these matrix pro-teins in a concentration-dependent manner, which is not the case for RGD-independent cell adhesion molecules like laminin and collagen [57]. Lactoferrin does not affect attachment of human dermal fibroblasts on cell culture substrate. The differential effect of lactoferrin on cell adhesion requires further study.

On the other hand, the gene expression of matrix metalloprotease-1 (MMP-1) is activated by lactoferrin in fibroblasts [58]. MMPs cleave a specific set of ECM pro-teins, and play important roles in wound healing, including facilitating migration of cells, removing damaged ECM and remodelling new matrix. MMP-1 cleaves intact fibrillar type I collagen, the predominant form of collagen in skin, at a single site in the collagen molecule. Therefore, it is possible that lactoferrin modulates cell motil-ity by regulating MMP-1 expression.

5.6   Hyaluronan Synthesis

Hyaluronan is a major component of the extracellular matrix (ECM) in dermis and epidermis. Hyaluronan is implicated in a number of wound healing processes, including cell migration, proliferation, and moderation of the inflammatory response [59]. Hyaluronan level is transiently increased in granulation tissue during the wound healing process. It is critical for formation of granulation tissue. Hyaluronan makes space that allows the infiltration of neutrophils and macrophages in the plasma clot. In granulation tissue, hyaluronan can interact with collagen, fibronectin and fibrinogen, and promotes fibrin polymerization and clot formation. Hyaluronan accumulation is essential for TGF-β-induced myofibroblast differentiation that plays important roles in wound contraction. Bovine lactoferrin increases the intracellular hyaluronan level in normal human dermal fibroblasts [60]. The accumulation of hyaluronan in conditioned media is also increased by lactoferrin. The elevation of hyaluronan is accompanied by elevation of *HAS2* (hyaluronan synthase 2) mRNA transcription and HAS2 protein expression [60] whereas *HAS1* mRNA transcription and HAS1 protein expression are not significantly increased by bovine lactoferrin. The promoting effects of bovine lactoferrin on hyaluronan production and HAS2 expression are inhibited by SB431542, a potent and selective inhibitor of the TGF-β receptor-1, suggesting that the wound healing activity of lactoferrin is due, in part, to the promotion of hyaluronan synthesis in dermal fibroblasts.

5.7   Production of Extracellular Matrix Components

Collagen deposition enhances granulation tissue formation and re-epithelialization in the skin-wound repair process. Fibroblasts from chronic wounds produce less collagen than normal dermal fibroblasts at least *in vitro* [61]. Both a reduction in collagen synthesis and in extracellular matrix stability could be observed in diabetes. Lactoferrin increases *COL1A1* mRNA expression and collagen biosynthesis by human dermal fibroblasts [60]. Hyaluronan itself promotes collagen synthesis in fetal rabbit dermal fibroblasts [62], and is therefore likely to support wound healing by providing ECM that promotes fibroblast migration. This property of lactoferrin accelerates the wound closure in normal and chronic wound.

5.8   Corneal Epithelia Wound Healing

Chemical injuries to corneal epithelium induce acute inflammation, characterized by rapid infiltration of neutrophils into wounded area. If treatment is delayed, corneal damage causes chronic inflammation accompanied by corneal vascularization and recruitment of macrophages and lymphocytes over extended periods, which followed by formation of recurrent corneal erosions, corneal ulceration, perforation, scar formation and permanent loss of vision. Prompt corneal healing is required to prevent chronic inflammation and to maintain a transparent functional cornea. Growth factors and cytokines play important roles in the corneal wound healing. PDGF, epidermal growth factor (EGF) and keratinocyte growth factor (KGF) accelerate the corneal epithelial wound healing [63–65]. Expressions of IL-1α and IL-6 are strongly induced after corneal injury [66]. IL-6 promotes corneal epithelial cell migration and wound closure *in vivo* by fibronectin dependent mechanism [67].

Bovine lactoferrin promotes the wound closure of alkaline injured HCLC in a dose-dependent manner [13]. Furthermore, lactoferrin treatment results in accelerated wound closure in mouse cornea alkaline wound model [13]. The effect of bovine lactoferrin C-lobe is more prominent than either the N-lobe or intact bovine lactoferrin [68]. Lactoferrin promotes PDGF and IL-6 production in both un-wounded and wounded HCLC. Both AG1295 (an PDGF receptor tyrosine kinase inhibitor) and IL-6 receptor neutralizing antibody inhibit the promoting effect of lactoferrin on wound closure in HCLC, suggesting their involvement in lactoferrin-enhanced wound closure [13]. Lactoferrin also regulates expression of pro-inflammatory and anti-inflammatory cytokines in the process of corneal wound closure. Lactoferrin induces slight increase of IL-1 and tumor necrosis factor-α (TNF-α). IL-1 is essential for early phase of alkaline wound closure [13]. The absence of TNF-α causes delay of corneal wound closure. Lactoferrin treatment results in down-regulation of IL-8, a strong chemotaxis factor for neutrophils, and recruits the polymorphonuclear (PMN) leukocytes at the site of corneal injury and inflammation.

5.9 Talactoferrin

Talactoferrin is a recombinant human lactoferrin produced in *Aspergillus awamori* [69]. Despite of a different glycosylation pattern, talactoferrin has been shown to retain biological activities including cancer prevention, anti-microbial and anti-inflammatory properties [70, 71]. Topical administration of talactoferrin enhances the rate of wound closure in control and diabetic db-/db- mice [15]. The effect of talactoferrin is more prominent than recombinant human PDGF. Wounds treated with a talactoferrin show increased production of IL-6, macrophage inflammatory protein-1 (MIP-1), MIP-2 (mouse IL-8 homologue) and TNF-α *in vivo*. Lactoferrin also promotes the production of IL-6, IL-8 and monocyte chemotactic protein 1 (MCP-1) in human dermal fibroblasts [15]. These are known as key inflammatory regulators that have been shown to peak early in wound models. IL-6 is produced by inflammatory and resident cells and has a crucial role in the pathogenesis of various inflammations. The reduction of the wound area is delayed in IL-6-knockout mice in which attenuated leukocyte infiltration, re-epithelialization, angiogenesis, and collagen accumulation has been observed [72]. The production of IL-8, MCP-1 and MIP-1α is increased in the wound healing process [72]. These cytokines play an indispensable biological role in the inflammatory phase of wound healing, especially in the recruitment of inflammatory leukocytes. The results of receptor competition binding assays indicate that talactoferrin binds IL-8 receptor B (IL-8 RB/CXCR2) and CCR2 [15]. These observations suggest that talactoferrin promotes wound healing by promoting early inflammatory phase of wound healing.

5.10 Cutaneous Immunity

Administration of a solvent antigen such as oxazolone induces migration of the Langerhans cells from the epidermis to the draining lymph nodes [73]. The accumulation of Langerhans cells is important for primary immune response since Langerhans cells act as antigen-presenting dendritic cells in lymph nodes. Following the administration of allergen to the skin surface, the release of IL-1β from epidermal Langerhans cells and TNF-α from keratinocytes is enhanced. IL-1β and TNF-α can induce the migration of Langerhans cells in lympho nodes [74, 75].

Lactoferrin is released from neutrophils in response to skin allergenic reactions [76]. Both intradermal injection and topical administration of lactoferrin can deliver lactoferrin to epidermal cells. Administration of lactoferrin by either route inhibits oxazolone-induced migration of Langerhans cells and the accumulation of dendritic cells within lymph nodes [77]. The inhibitory effect of lactoferrin is independent from its iron saturation statues. Lactoferrin inhibits the IL-1β-induced migration of Langerhans cells, indicating that lactoferrin can directly suppress the cutaneous inflammatory response in response to injury *in vivo*.

5.11 Anti-apoptotic Effect

Apoptosis is increased in a model of diabetes-impaired wound healing (chronic wound) in non-obese diabetic (NOD) mice [78]. In chronic wound, expression of growth factors and their receptor is decreased [79, 80]. Consequently, the fibroblasts show decreased response to the growth factors, results in their increased apoptosis and delayed wound repair [78]. Current topically applied growth factors do not induce cell proliferation in chronic wounds. Rice derived recombinant human lactoferrin prevents apoptosis of human dermal fibroblasts induced by serum deprivation or TPA (12-O-tetradecanoylphorbol-13-acetate) exposure [16]. Furthermore, the anti-apoptotic effect of lactoferrin could be observed for human epidermal keratinocytes [17]. These observations suggest that lactoferrin promotes the chronic wound healing by preventing loss of fibroblasts and keratinocytes.

5.12 Diabetic Ulcers

Ulcer is one of the most common and most devastating complication of diabetes. A special feature of diabetic chronic wound is prolonged or excessive inflammatory response at the wounded site. It is associated with peripheral neuropathy and reduced arterial blood flow. Ulcers contain excessive ROS that further damage the cells and the healing tissues [81]. Furthermore, delayed wound healing increases the chance of bacterial infections. Particularly, chronic wound is prone to bacterial infection, often leading to more serious complications such as gangrene and amputation [82].

In a clinical trial of patients with diabetic foot ulcers, the talactoferrin treated group shows a higher wound healing rates compared with placebo group [83]. Topical administration of talactoferrin enhances the rate of wound closure in control and diabetic db⁻/db⁻ mice [15]. Lactoferrin is known as its anti-bacterial, anti-viral activities which may contribute to the healing of diabetic ulcers. Actually, biofilm formation is prevented in wounds treated with lactoferrin in combination with xylitol [84]. In addition, lactoferrin is thought to exhibit anti-inflammatory activity, thus neutralizing an overabundant immune response. Lactoferrin acts as an iron scavenger, and thus prevents the oxidative damage to tissues by inhibiting the ROS production. These activities of lactoferrin may contribute the care of chronic diabetic ulcers. According to a proteomic analysis, lactoferrin is identified as protein having the highest abundance in exudates obtained from chronic wound compared with normal wounds [85]. Detailed analysis of lactoferrin function in chronic wound is required.

References

1. Martin P (1997) Wound healing–aiming for perfect skin regeneration. Science 276(5309): 75–81
2. Eming SA, Krieg T, Davidson JM (2007) Inflammation in wound repair: molecular and cellular mechanisms. J Invest Dermatol 127(3):514–525
3. Shaw TJ, Martin P (2009) Wound repair at a glance. J Cell Sci 122(Pt 18):3209–3213
4. Theilgaard-Monch K, Knudsen S, Follin P, Borregaard N (2004) The transcriptional activation program of human neutrophils in skin lesions supports their important role in wound healing. J Immunol 172(12):7684–7693
5. Conneely OM (2001) Antiinflammatory activities of lactoferrin. J Am Coll Nutr 20(5 Suppl): 389S–395S, discussion 396S–397S
6. Legrand D, Elass E, Carpentier M, Mazurier J (2005) Lactoferrin: a modulator of immune and inflammatory responses. Cell Mol Life Sci 62(22):2549–2559
7. Actor JK, Hwang SA, Kruzel ML (2009) Lactoferrin as a natural immune modulator. Curr Pharm Des 15(17):1956–1973
8. Ward PP, Uribe-Luna S, Conneely OM (2002) Lactoferrin and host defense. Biochem Cell Biol 80(1):95–102
9. Dovi JV, Szpaderska AM, DiPietro LA (2004) Neutrophil function in the healing wound: adding insult to injury? Thromb Haemost 92(2):275–280
10. Loots MA, Lamme EN, Zeegelaar J, Mekkes JR et al (1998) Differences in cellular infiltrate and extracellular matrix of chronic diabetic and venous ulcers versus acute wounds. J Invest Dermatol 111(5):850–857
11. Sivamani RK, Lam ST, Isseroff RR (2007) Beta adrenergic receptors in keratinocytes. Dermatol Clin 25(4):643–653, x
12. Nakajima M, Shinoda I, Samejima Y, Miyauchi H et al (1997) Lactoferrin as a suppressor of cell migration of gastrointestinal cell lines. J Cell Physiol 170(2):101–105
13. Pattamatta U, Willcox M, Stapleton F, Cole N et al (2009) Bovine lactoferrin stimulates human corneal epithelial alkali wound healing in vitro. Invest Ophthalmol Vis Sci 50(4): 1636–1643

14. Takayama Y, Mizumachi K (2001) Effects of lactoferrin on collagen gel contractile activity and myosin light chain phosphorylation in human fibroblasts. FEBS Lett 508(1):111–116
15. Engelmayer J, Blezinger P, Varadhachary A (2008) Talactoferrin stimulates wound healing with modulation of inflammation. J Surg Res 149(2):278–286
16. Tang L, Cui T, Wu JJ, Liu-Mares W et al (2010) A rice-derived recombinant human lactoferrin stimulates fibroblast proliferation, migration, and sustains cell survival. Wound Repair Regen 18(1):123–131
17. Tang L, Wu JJ, Ma Q, Cui T et al (2010) Human lactoferrin stimulates skin keratinocyte function and wound re-epithelialization. Br J Dermatol 163(1):38–47
18. Morasso MI, Tomic-Canic M (2005) Epidermal stem cells: the cradle of epidermal determination, differentiation and wound healing. Biol Cell 97(3):173–183
19. Grinnell F (1994) Fibroblasts, myofibroblasts, and wound contraction. J Cell Biol 124(4): 401–404
20. Grinnell F, Ho CH, Lin YC, Skuta G (1999) Differences in the regulation of fibroblast contraction of floating versus stressed collagen matrices. J Biol Chem 274(2):918–923
21. Takayama Y, Mizumachi K, Takezawa T (2002) The bovine lactoferrin region responsible for promoting the collagen gel contractile activity of human fibroblasts. Biochem Biophys Res Commun 299(5):813–817
22. Pellegrin S, Mellor H (2007) Actin stress fibres. J Cell Sci 120(Pt 20):3491–3499
23. Totsukawa G, Yamakita Y, Yamashiro S, Hartshorne DJ et al (2000) Distinct roles of ROCK (Rho-kinase) and MLCK in spatial regulation of MLC phosphorylation for assembly of stress fibers and focal adhesions in 3 T3 fibroblasts. J Cell Biol 150(4):797–806
24. Totsukawa G, Wu Y, Sasaki Y, Hartshorne DJ et al (2004) Distinct roles of MLCK and ROCK in the regulation of membrane protrusions and focal adhesion dynamics during cell migration of fibroblasts. J Cell Biol 164(3):427–439
25. Webb DJ, Donais K, Whitmore LA, Thomas SM et al (2004) FAK-Src signalling through paxillin, ERK and MLCK regulates adhesion disassembly. Nat Cell Biol 6(2):154–161
26. Huang C, Jacobson K, Schaller MD (2004) MAP kinases and cell migration. J Cell Sci 117 (Pt 20):4619–4628
27. Klemke RL, Cai S, Giannini AL, Gallagher PJ et al (1997) Regulation of cell motility by mitogen-activated protein kinase. J Cell Biol 137(2):481–492
28. Nguyen DH, Catling AD, Webb DJ, Sankovic M et al (1999) Myosin light chain kinase functions downstream of Ras/ERK to promote migration of urokinase-type plasminogen activator-stimulated cells in an integrin-selective manner. J Cell Biol 146(1):149–164
29. Takayama Y, Takahashi H, Mizumachi K, Takezawa T (2003) Low density lipoprotein receptor-related protein (LRP) is required for lactoferrin-enhanced collagen gel contractile activity of human fibroblasts. J Biol Chem 278(24):22112–22118
30. Somlyo AP, Somlyo AV (2000) Signal transduction by G-proteins, rho-kinase and protein phosphatase to smooth muscle and non-muscle myosin II. J Physiol 522(Pt 2):177–185
31. Parizi M, Howard EW, Tomasek JJ (2000) Regulation of LPA-promoted myofibroblast contraction: role of Rho, myosin light chain kinase, and myosin light chain phosphatase. Exp Cell Res 254(2):210–220
32. Amano M, Mukai H, Ono Y, Chihara K et al (1996) Identification of a putative target for Rho as the serine-threonine kinase protein kinase N. Science 271(5249):648–650
33. Vash B, Phung N, Zein S, DeCamp D (1998) Three complement-type repeats of the low-density lipoprotein receptor-related protein define a common binding site for RAP, PAI-1, and lactoferrin. Blood 92(9):3277–3285
34. Neels JG, van Den Berg BM, Lookene A, Olivecrona G et al (1999) The second and fourth cluster of class A cysteine-rich repeats of the low density lipoprotein receptor-related protein share ligand-binding properties. J Biol Chem 274(44):31305–31311
35. May P, Herz J, Bock HH (2005) Molecular mechanisms of lipoprotein receptor signalling. Cell Mol Life Sci 62(19–20):2325–2338
36. May P, Woldt E, Matz RL, Boucher P (2007) The LDL receptor-related protein (LRP) family: an old family of proteins with new physiological functions. Ann Med 39(3):219–228

37. Lillis AP, Mikhailenko I, Strickland DK (2005) Beyond endocytosis: LRP function in cell migration, proliferation and vascular permeability. J Thromb Haemost 3(8):1884–1893

38. Okada SS, Grobmyer SR, Barnathan ES (1996) Contrasting effects of plasminogen activators, urokinase receptor, and LDL receptor-related protein on smooth muscle cell migration and invasion. Arterioscler Thromb Vasc Biol 16(10):1269–1276

39. Cheng CF, Fan J, Fedesco M, Guan S et al (2008) Transforming growth factor alpha (TGFalpha)-stimulated secretion of HSP90alpha: using the receptor LRP-1/CD91 to promote human skin cell migration against a TGFbeta-rich environment during wound healing. Mol Cell Biol 28(10):3344–3358

40. Nassar T, Haj-Yehia A, Akkawi S, Kuo A et al (2002) Binding of urokinase to low density lipoprotein-related receptor (LRP) regulates vascular smooth muscle cell contraction. J Biol Chem 277(43):40499–40504

41. Sumi Y, Muramatsu H, Hata K, Ueda M et al (2000) Midkine enhances early stages of collagen gel contraction. J Biochem 127(2):247–251

42. Muramatsu H, Zou K, Sakaguchi N, Ikematsu S et al (2000) LDL receptor-related protein as a component of the midkine receptor. Biochem Biophys Res Commun 270(3):936–941

43. Willnow TE, Goldstein JL, Orth K, Brown MS et al (1992) Low density lipoprotein receptor-related protein and gp330 bind similar ligands, including plasminogen activator-inhibitor complexes and lactoferrin, an inhibitor of chylomicron remnant clearance. J Biol Chem 267(36): 26172–26180

44. Ji ZS, Mahley RW (1994) Lactoferrin binding to heparan sulfate proteoglycans and the LDL receptor-related protein. Further evidence supporting the importance of direct binding of remnant lipoproteins to HSPG. Arterioscler Thromb 14(12):2025–2031

45. Meilinger M, Haumer M, Szakmary KA, Steinbock F et al (1995) Removal of lactoferrin from plasma is mediated by binding to low density lipoprotein receptor-related protein/alpha 2-macroglobulin receptor and transport to endosomes. FEBS Lett 360(1):70–74

46. Fillebeen C, Descamps L, Dehouck MP, Fenart L et al (1999) Receptor-mediated transcytosis of lactoferrin through the blood-brain barrier. J Biol Chem 274(11):7011–7017

47. Goretzki L, Mueller BM (1998) Low-density-lipoprotein-receptor-related protein (LRP) interacts with a GTP-binding protein. Biochem J 336(Pt 2):381–386

48. Bu G, Marzolo MP (2000) Role of rap in the biogenesis of lipoprotein receptors. Trends Cardiovasc Med 10(4):148–155

49. Birkenmeier G, Heidrich K, Glaser C, Handschug K et al (1998) Different expression of the alpha2-macroglobulin receptor/low-density lipoprotein receptor-related protein in human keratinocytes and fibroblasts. Arch Dermatol Res 290(10):561–568

50. Grey A, Banovic T, Zhu Q, Watson M et al (2004) The low-density lipoprotein receptor-related protein 1 is a mitogenic receptor for lactoferrin in osteoblastic cells. Mol Endocrinol 18(9): 2268–2278

51. Dhennin-Duthille I, Masson M, Damiens E, Fillebeen C et al (2000) Lactoferrin upregulates the expression of CD4 antigen through the stimulation of the mitogen-activated protein kinase in the human lymphoblastic T Jurkat cell line. J Cell Biochem 79(4):583–593

52. Gotthardt M, Trommsdorff M, Nevitt MF, Shelton J et al (2000) Interactions of the low density lipoprotein receptor gene family with cytosolic adaptor and scaffold proteins suggest diverse biological functions in cellular communication and signal transduction. J Biol Chem 275(33): 25616–25624

53. Loukinova E, Ranganathan S, Kuznetsov S, Gorlatova N et al (2002) Platelet-derived growth factor (PDGF)-induced tyrosine phosphorylation of the low density lipoprotein receptor-related protein (LRP). Evidence for integrated co-receptor function betwenn LRP and the PDGF. J Biol Chem 277(18):15499–15506

54. Barnes H, Ackermann EJ, van der Geer P (2003) v-Src induces Shc binding to tyrosine 63 in the cytoplasmic domain of the LDL receptor-related protein 1. Oncogene 22(23):3589–3597

55. Shinmoto H, Sato K, Dosako S (1992) Inhibition by bovine lactoferrin of adhesion of L929 cells cultured in serum-free git medium. Biosci Biotechnol Biochem 56(6):965–966

56. Pollanen MT, Hakkinen L, Overman DO, Salonen JI (1998) Lactoferrin impedes epithelial cell adhesion in vitro. J Periodontal Res 33(1):8–16

57. Sakamoto K, Ito Y, Mori T, Sugimura K (2006) Interaction of human lactoferrin with cell adhesion molecules through RGD motif elucidated by lactoferrin-binding epitopes. J Biol Chem 281(34):24472–24478

58. Oh SM, Hahm DH, Kim IH, Choi SY (2001) Human neutrophil lactoferrin trans-activates the matrix metalloproteinase 1 gene through stress-activated MAPK signaling modules. J Biol Chem 276(45):42575–42579

59. Stern R (2003) Devising a pathway for hyaluronan catabolism: are we there yet? Glycobiology 13(12):105R–115R

60. Saito S, Takayama Y, Mizumachi K, Suzuki C (2011) Lactoferrin promotes hyaluronan synthesis in human dermal fibroblasts. Biotechnol Lett 33(1):33–39

61. Herrick SE, Ireland GW, Simon D, McCollum CN et al (1996) Venous ulcer fibroblasts compared with normal fibroblasts show differences in collagen but not fibronectin production under both normal and hypoxic conditions. J Invest Dermatol 106(1):187–193

62. Mast BA, Diegelmann RF, Krummel TM, Cohen IK (1993) Hyaluronic acid modulates proliferation, collagen and protein synthesis of cultured fetal fibroblasts. Matrix 13(6):441–446

63. Lu L, Reinach PS, Kao WWY (2001) Corneal epithelial wound healing. Exp Biol Med 226(7):653–664

64. Yu FS, Yin J, Xu K, Huang J (2010) Growth factors and corneal epithelial wound healing. Brain Res Bull 81(2–3):229–235

65. Imanishi J, Kamiyama K, Iguchi I, Kita M et al (2000) Growth factors: importance in wound healing and maintenance of transparency of the cornea. Prog Retin Eye Res 19(1):113–129

66. Sotozono C, He JC, Matsumoto Y, Kita M et al (1997) Cytokine expression in the alkali-burned cornea. Curr Eye Res 16(7):670–676

67. Nishida T, Nakamura M, Mishima H, Otori T (1992) Interleukin 6 promotes epithelial migration by a fibronectin-dependent mechanism. J Cell Physiol 153(1):1–5

68. Ashby B, Garrett Q, Willcox M (2011) Bovine lactoferrin structures promoting corneal epithelial wound healing in vitro. Invest Ophthalmol Vis Sci 52(5):2719–2726

69. Ward PP, Piddington CS, Cunningham GA, Zhou X et al (1995) A system for production of commercial quantities of human lactoferrin: a broad spectrum natural antibiotic. Biotechnology (NY) 13(5):498–503

70. Hayes TG, Falchook GF, Varadhachary GR, Smith DP et al (2006) Phase I trial of oral talactoferrin alfa in refractory solid tumors. Invest New Drugs 24(3):233–240

71. Spadaro M, Caorsi C, Ceruti P, Varadhachary A et al (2008) Lactoferrin, a major defense protein of innate immunity, is a novel maturation factor for human dendritic cells. FASEB J 22(8):2747–2757

72. Lin ZQ, Kondo T, Ishida Y, Takayasu T et al (2003) Essential involvement of IL-6 in the skin wound-healing process as evidenced by delayed wound healing in IL-6-deficient mice. J Leukoc Biol 73(6):713–721

73. Kinnaird A, Peters SW, Foster JR, Kimber I (1989) Dendritic cell accumulation in draining lymph nodes during the induction phase of contact allergy in mice. Int Arch Allergy Appl Immunol 89(2–3):202–210

74. Cumberbatch M, Kimber I (1992) Dermal tumour necrosis factor-alpha induces dendritic cell migration to draining lymph nodes, and possibly provides one stimulus for Langerhans' cell migration. Immunology 75(2):257–263

75. Cumberbatch M, Dearman RJ, Kimber I (1997) Langerhans cells require signals from both tumour necrosis factor alpha and interleukin 1 beta for migration. Adv Exp Med Biol 417:125–128

76. Zweiman B, Kucich U, Shalit M, Von Allmen C et al (1990) Release of lactoferrin and elastase in human allergic skin reactions. J Immunol 144(10):3953–3960

77. Cumberbatch M, Dearman RJ, Uribe-Luna S, Headon DR et al (2000) Regulation of epidermal Langerhans cell migration by lactoferrin. Immunology 100(1):21–28

78. Darby IA, Bisucci T, Hewitson TD, MacLellan DG (1997) Apoptosis is increased in a model of diabetes-impaired wound healing in genetically diabetic mice. Int J Biochem Cell Biol 29(1):191–200
79. Robson MC (1997) The role of growth factors in the healing of chronic wounds. Wound Repair Regen 5(1):12–17
80. Kim HM, Lowery JC, Hamill JB, Wilkins EG (2003) Accuracy of a web-based system for monitoring chronic wounds. Telemed J E Health 9(2):129–140
81. Soneja A, Drews M, Malinski T (2005) Role of nitric oxide, nitroxidative and oxidative stress in wound healing. Pharmacol Rep 57(Suppl):108–119
82. LeFrock JL, Joseph WS (1995) Bone and soft-tissue infections of the lower extremity in diabetics. Clin Podiatr Med Surg 12(1):87–103
83. Lyons TE, Miller MS, Serena T, Sheehan P et al (2007) Talactoferrin alfa, a recombinant human lactoferrin promotes healing of diabetic neuropathic ulcers: a phase 1/2 clinical study. Am J Surg 193(1):49–54
84. Ammons MC, Ward LS, James GA (2011) Anti-biofilm efficacy of a lactoferrin/xylitol wound hydrogel used in combination with silver wound dressings. Int Wound J 8(3):268–273
85. Eming SA, Koch M, Krieger A, Brachvogel B et al (2010) Differential proteomic analysis distinguishes tissue repair biomarker signatures in wound exudates obtained from normal healing and chronic wounds. J Proteome Res 9(9):4758–4766

Index